Plant Hormone Protocols

METHODS IN MOLECULAR BIOLOGY™

John M. Walker, Series Editor

141. **Plant Hormone Protocols**, edited by *Gregory A. Tucker and Jeremy A. Roberts, 2000*
140. **Chaperonin Protocols**, edited by *Christine Schneider, 2000*
139. **Extracellular Matrix Protocols**, edited by *Charles Streuli and Michael Grant, 2000*
138. **Chemokine Protocols**, edited by *Amanda E. I. Proudfoot, Timothy N. C. Wells, and Christine Power, 2000*
137. **Developmental Biology Protocols, Volume III**, edited by *Rocky S. Tuan and Cecilia W. Lo, 2000*
136. **Developmental Biology Protocols, Volume II**, edited by *Rocky S. Tuan and Cecilia W. Lo, 2000*
135. **Developmental Biology Protocols, Volume I**, edited by *Rocky S. Tuan and Cecilia W. Lo, 2000*
134. **T Cell Protocols:** *Development and Activation*, edited by *Kelly P. Kearse, 2000*
133. **Gene Targeting Protocols**, edited by *Eric B. Kmiec, 2000*
132. **Bioinformatics Methods and Protocols**, edited by *Stephen Misener and Stephen A. Krawetz, 2000*
131. **Flavoprotein Protocols**, edited by *S. K. Chapman and G. A. Reid, 1999*
130. **Transcription Factor Protocols**, edited by *Martin J. Tymms, 2000*
129. **Integrin Protocols**, edited by *Anthony Howlett, 1999*
128. **NMDA Protocols**, edited by *Min Li, 1999*
127. **Molecular Methods in Developmental Biology:** Xenopus *and Zebrafish*, edited by *Matthew Guille, 1999*
126. **Adrenergic Receptor Protocols**, edited by *Curtis A. Machida, 2000*
125. **Glycoprotein Methods and Protocols:** *The Mucins*, edited by *Anthony P. Corfield, 2000*
124. **Protein Kinase Protocols**, edited by *Alastair D. Reith, 2000*
123. ***In Situ* Hybridization Protocols (2nd ed.)**, edited by *Ian A. Darby, 2000*
122. **Confocal Microscopy Methods and Protocols**, edited by *Stephen W. Paddock, 1999*
121. **Natural Killer Cell Protocols:** *Cellular and Molecular Methods*, edited by *Kerry S. Campbell and Marco Colonna, 2000*
120. **Eicosanoid Protocols**, edited by *Elias A. Lianos, 1999*
119. **Chromatin Protocols**, edited by *Peter B. Becker, 1999*
118. **RNA–Protein Interaction Protocols**, edited by *Susan R. Haynes, 1999*
117. **Electron Microscopy Methods and Protocols**, edited by *M. A. Nasser Hajibagheri, 1999*
116. **Protein Lipidation Protocols**, edited by *Michael H. Gelb, 1999*
115. **Immunocytochemical Methods and Protocols (2nd ed.)**, edited by *Lorette C. Javois, 1999*
114. **Calcium Signaling Protocols**, edited by *David G. Lambert, 1999*
113. **DNA Repair Protocols:** *Eukaryotic Systems*, edited by *Daryl S. Henderson, 1999*
112. **2-D Proteome Analysis Protocols**, edited by *Andrew J. Link, 1999*
111. **Plant Cell Culture Protocols**, edited by *Robert D. Hall, 1999*
110. **Lipoprotein Protocols**, edited by *Jose M. Ordovas, 1998*
109. **Lipase and Phospholipase Protocols**, edited by *Mark H. Doolittle and Karen Reue, 1999*
108. **Free Radical and Antioxidant Protocols**, edited by *Donald Armstrong, 1998*
107. **Cytochrome P450 Protocols**, edited by *Ian R. Phillips and Elizabeth A. Shephard, 1998*
106. **Receptor Binding Techniques**, edited by *Mary Keen, 1999*
105. **Phospholipid Signaling Protocols**, edited by *Ian M. Bird, 1998*
104. **Mycoplasma Protocols**, edited by *Roger J. Miles and Robin A. J. Nicholas, 1998*
103. ***Pichia* Protocols**, edited by *David R. Higgins and James M. Cregg, 1998*
102. **Bioluminescence Methods and Protocols**, edited by *Robert A. LaRossa, 1998*
101. **Mycobacteria Protocols**, edited by *Tanya Parish and Neil G. Stoker, 1998*
100. **Nitric Oxide Protocols**, edited by *Michael A. Titheradge, 1998*
99. **Stress Response:** *Methods and Protocols*, edited by *Stephen M. Keyse, 2000*
98. **Forensic DNA Profiling Protocols**, edited by *Patrick J. Lincoln and James M. Thomson, 1998*
97. **Molecular Embryology:** *Methods and Protocols*, edited by *Paul T. Sharpe and Ivor Mason, 1999*
96. **Adhesion Protein Protocols**, edited by *Elisabetta Dejana and Monica Corada, 1999*
95. **DNA Topoisomerases Protocols:** *II. Enzymology and Drugs*, edited by *Mary-Ann Bjornsti and Neil Osheroff, 2000*
94. **DNA Topoisomerases Protocols:** *I. DNA Topology and Enzymes*, edited by *Mary-Ann Bjornsti and Neil Osheroff, 1999*
93. **Protein Phosphatase Protocols**, edited by *John W. Ludlow, 1998*
92. **PCR in Bioanalysis**, edited by *Stephen J. Meltzer, 1998*
91. **Flow Cytometry Protocols**, edited by *Mark J. Jaroszeski, Richard Heller, and Richard Gilbert, 1998*
90. **Drug–DNA Interaction Protocols**, edited by *Keith R. Fox, 1998*
89. **Retinoid Protocols**, edited by *Christopher Redfern, 1998*
88. **Protein Targeting Protocols**, edited by *Roger A. Clegg, 1998*
87. **Combinatorial Peptide Library Protocols**, edited by *Shmuel Cabilly, 1998*
86. **RNA Isolation and Characterization Protocols**, edited by *Ralph Rapley and David L. Manning, 1998*
85. **Differential Display Methods and Protocols**, edited by *Peng Liang and Arthur B. Pardee, 1997*
84. **Transmembrane Signaling Protocols**, edited by *Dafna Bar-Sagi, 1998*
83. **Receptor Signal Transduction Protocols**, edited by *R. A. John Challiss, 1997*
82. **Arabidopsis Protocols**, edited by *José M Martinez-Zapater and Julio Salinas, 1998*
81. **Plant Virology Protocols:** *From Virus Isolation to Transgenic Resistance*, edited by *Gary D. Foster and Sally Taylor, 1998*
80. **Immunochemical Protocols (2nd. ed.)**, edited by *John Pound, 1998*
79. **Polyamine Protocols**, edited by *David M. L. Morgan, 1998*
78. **Antibacterial Peptide Protocols**, edited by *William M. Shafer, 1997*
77. **Protein Synthesis:** *Methods and Protocols*, edited by *Robin Martin, 1998*
76. **Glycoanalysis Protocols (2nd. ed.)**, edited by *Elizabeth F. Hounsell, 1998*

METHODS IN MOLECULAR BIOLOGY™

Plant Hormone Protocols

Edited by

Gregory A. Tucker

and

Jeremy A. Roberts

School of Biological Sciences, University of Nottingham, UK

Humana Press Totowa, New Jersey

999 Riverview Drive, Suite 208
Totowa, New Jersey 07512

This publication is printed on acid-free paper. ∞
ANSI Z39.48-1984 (American Standards Institute) Permanence of Paper for Printed Library Materials.

Cover design by Patricia F. Cleary.

For additional copies, pricing for bulk purchases, and/or information about other Humana titles, contact Humana at the above address or at any of the following numbers: Tel.: 973-256-1699; Fax: 973-256-8341; E-mail: humana@humanapr.com; Website: http://humanapress.com

Printed in the United States of America. 10 9 8 7 6 5 4 3 2 1

Library of Congress Cataloging in Publication Data

Plant hormone protocols / Gregory A. Tucker and Jeremy A. Roberts.
p. cm. -- (Methods in molecular biology ; v. 141)
Includes bibliographical references (p.).
ISBN 0-89603-577-8 (alk. paper)
1. Plant hormones. I. Tucker, G. A. (Gregory A.) II. Roberts, J. A. (Jeremy A.) III. Series.

QK898.H67 P63 2000
571.7'42--dc21

99-058889

Preface

Plant scientists have been aware for over 50 years that minute amounts of a group of compounds termed hormones or growth regulators can have a profound effect on differentiation and development. As the years have gone by, increasingly sophisticated approaches have been developed to determine the concentrations of these chemicals in plant tissues and cells, and *Plant Hormone Protocols* seeks to provide a "state-of-the-art" account of these techniques.

What is clear is that in addition to knowing the amounts of plant hormone molecules, we also need to know how they are produced and how they bring about their effects. Consequently, we have also included contemporary approaches for the identification of the biosynthetic pathways of plant hormones, the receptors with which they interact, and the possible signaling systems by which they exert their effects. We have not aimed to cover these different aspects with every plant hormone, but rather have chosen particularly powerful strategies that we believe will be appropriate in our quest to understand how the concentrations of these compounds are regulated and how they may bring about some of the responses that are observed.

We are confident that with the advent of new approaches in molecular biology and transgenic technology the next five years will see an explosion in information relating to these aspects of plant science.

Gregory A. Tucker
Jeremy A. Roberts

Contents

Contributors

SUZANNE R. ABRAMS • *Plant Biotechnology Institute, National Research Council of Canada, Saskatoon, Canada*

PETE BASS • *School of Biological Sciences, University of Nottingham, UK*

MALCOLM J. BENNETT • *School of Biological Sciences, University of Nottingham, UK*

CORNELIUS W. P. M. BLOM • *Department of Ecology, University of Nijmegen, Nijmegen, The Netherlands.*

NICHOLAS J. BREWIN • *Department of Genetics, John Innes Centre, Norwich Research Park, Colney, Norwich, UK*

DEBORAH CLEMENTS • *School of Biological Sciences, University of Nottingham, UK*

STEPHEN J. CROKER • *Department of Agricultural Sciences, IARC-Long Ashton Research Station, University of Bristol, Long Ashton, Bristol, UK*

HUGO S. M. DE VRIES • *ATO-DLO, Wageningen, The Netherlands*

BJØRN K. DRØBAK • *Cell Signalling Group, Department of Cell Biology, John Innes Centre, Norwich Research Park, Colney, Norwich, UK*

FRANS J. M. HARREN • *Department of Molecular and Laser Physics, University of Nijmegen, Nijmegen, The Netherlands*

PETER HEDDEN • *Department of Agricultural Sciences, IARC-Long Ashton Research Station, University of Bristol, Long Ashton, Bristol, UK*

SACCO TE LINTEL HEKKERT • *Department of Molecular and Laser Physics, University of Nijmegen, Nijmegen, The Netherlands*

LUIS E. HERNANDEZ • *Laboratory of Plant Physiology, Edificio de Biológicas, Universidad Autónoma de Madrid, Madrid, Spain*

LAWRENCE R. HOGGE • *Plant Biotechnology Institute, National Research Council of Canada, Saskatoon, Canada*

PAULA E. JAMESON • *Institute of Molecular BioSciences, College of Sciences, Massey University, Palmerston North, New Zealand*

DAVID H. LEWIS • *New Zealand Institute for Coop and Food Research, Palmerston North, New Zealand*

OTTOLINE LEYSER • *The Plant Laboratory, Department of Biology, University of York, Heslington, UK*

GRANTLEY W. LYCETT • *School of Biological Sciences, University of Nottingham, UK*
SEAN T. MAY • *School of Biological Sciences, University of Nottingham, UK*
SEVGI ÖDEN • *Department of Biology, University of Antwerp (UIA), Antwerpen, Belgium*
ELS PRINSEN • *Department of Biology, University of Antwerp (UIA), Antwerpen, Belgium*
JEREMY A. ROBERTS • *School of Biological Sciences, University of Nottingham, UK*
PATRICIA A. ROSE • *Plant Biotechnology Institute, National Research Council of Canada, Saskatoon, Canada*
BENEDETTO RUPERTI • *School of Biological Sciences, University of Nottingham, UK*
COR A. SIKKENS • *Department of Molecular and Laser Physics, University of Nijmegen, Nijmegen, The Netherlands*
IAN TAYLOR • *School of Biological Sciences, University of Nottingham, UK*
GREGORY A. TUCKER • *School of Biological Sciences, University of Nottingham, UK*
STIJN VAN LAER • *Department of Biology, University of Antwerp (UIA), Antwerpen, Belgium*
HENRI VAN ONCKELEN • *Department of Biology, University of Antwerp (UIA), Antwerpen, Belgium*
MICHAEL A. VENIS • *Horticulture Research International, Wellesbourne, Warwick, UK*
LAURENTIUS A. C. J. VOESENEK • *Plant Ecophysiology, Faculty of Biology, Utrecht University, Utrecht, The Netherlands.*
M. K. WALKER-SIMMONS • *Agriculture Research Service, Washington State University, US Department of Agriculture, Pullman, WA*
CATHERINE A. WHITELAW • *Agriculture Research Service SARL, US Department of Agriculture, Beltsville, MD*
HUAIBI ZHANG • *The Horticulture and Food Research Institute of New Zealand, Palmerston North, New Zealand*

Plant Hormone Protocols

1

Extraction and Purification of an Enzyme Potentially Involved in ABA Biosynthesis

Gregory A. Tucker, Pete Bass, and Ian Taylor

1. Introduction

The advent of genetic engineering has provided a means to manipulate the biosynthesis of plant hormones to the advantage of agriculture. Such manipulation is very dependent on a detailed knowledge of the biosynthetic pathways involved in the production of the hormones, and more precisely about the enzymes involved in this biosynthesis. The knowledge of enzymes assists, but is not always a prerequisite for, the isolation of the genetic material, usually cDNAs, used as the genetic tools to manipulate hormone synthesis.

The first such example of the manipulation of hormone biosynthesis by the application of genetic engineering was for the hormone ethene. The biosynthetic pathway for ethene was first described in apples by Adams and Yang *(1)*. Methionine was found to be the precursor of this hormone in plant tissue and is converted via *S*-adenysyl-methionine (SAM) into the unique compound 1-amino-1-carboxyl cyclopropane (ACC). The conversion of SAM into ACC is carried out by a key biosynthetic enzyme, ACC synthase, and the subsequent conversion of ACC into ethene is carried out by a second enzyme, namely ACC oxidase. These two enzymes

From: *Methods in Molecular Biology, vol. 141: Plant Hormone Protocols*
Edited by: G. A. Tucker and J. A. Roberts

have been extensively studied and cDNAs for both identified, although by completely different methods. ACC oxidase cDNA was initially identified by a process of differential screening of a tomato-ripening cDNA library. This technique was used to identify cDNA clones from a library made from ripe tomato, which are exclusively expressed in a ripe fruit but not a green fruit. Since the biosynthesis of ethene was known to be ripening specific, it was postulated that at least some of any clones identified would encode biosynthetic enzymes. The technique identified several clones, one of which, pTOM13, appeared to have expression closely correlated with ethene synthesis. The use of this cDNA to transform tomato plants with an antisense gene resulted in genetically manipulated plants with lowered ethene production and a correspondingly reduced ability to convert ACC to ethene *(2)*. It was thus postulated that the pTOM13 cDNA encoded the enzyme ACC oxidase, a conclusion further supported by the observation that yeast genetically engineered with pTOM13 cDNA acquired the ability to convert ACC to ethene *(3)*. The sequence analysis of the pTOM 13 cDNA suggested a close relationship between the ACC oxidase and flavonone oxidase enzymes. This information enabled Ververdis and John *(4)* to successfully extract the ACC oxidase enzyme and carry out its purification and characterization.

The isolation of a cDNA encoding ACC synthase was achieved by a different route. In this instance the enzyme itself was first extracted and partially purified, and then an antibody was raised against the protein. Using this antibody a cDNA expression library was screened and the corresponding cDNA isolated *(5)*. The identity of the cDNA was confirmed by the expression of ACC synthase activity in *Escherichia coli* transformed with the cDNA *(5)*. This cDNA has been used to construct an antisense gene, which has in turn been used to silence members of the endogenous ACC synthase gene family and bring about a marked reduction in ethene synthesis in transgenic plants *(6)*. The resultant plants showed <0.5% of normal ethene production and since this was achieved in tomato, the fruit showed markedly inhibited ripening.

This technique of gene silencing could easily be used to reduce the synthesis of any or all plant hormones given sufficient knowledge of the biosynthetic pathways involved and the genetic tools, i.e., cDNAs, for key biosynthetic enzymes. The enzyme step(s) must be chosen with care to influence hormone synthesis exclusively. For instance, in the case of ethene the two target enzymes—ACC synthase and ACC oxidase—appear to be exclusively involved in the biosynthesis of this hormone and are not involved in any other key metabolic pathways. In contrast, had the enzyme converting methionine to SAM been the target for genetic manipulation, a more unpredictable phenotype may have arisen. This is because SAM is a key intermediate in several biosynthetic pathways, including those for cell walls (SAM provides the methyl groups for the esterification of pectin) and polyamines.

The general biosynthetic routes for several of the major plant hormones are fairly well established ***(7,8)***. Auxin (indole-3-acetic acid) has close structural similarity to tryptophan and tracer studies have identified tryptophan as a precursor for this auxin. However, the precise biosynthetic route is not clear. Several pathways have been postulated involving intermediates, such as indole-3-acetaldehyde and indole-3-acetamide ***(9)***. The gibberellins are all derived from the diterpenoid precursor geranyl–geranyl pyrophosphate, the over 70 naturally occuring gibberellins being the result of a complex branched biosynthetic pathway. The cytokinins are N-substituted adenines and as such could arise from the turnover of tRNA species or directly from adenine metabolism ***(10)***.

In all these cases cell-free systems have been reported that may retain at least part of the biosynthetic chain in an active state. Thus, for auxin, cell-free extracts have been demonstrated that can convert tryptophan to indole-3-acetic acid via the intermediates indole-3-pyruvic acid and indole-3-acetaldehyde. For gibberellins an enzyme—ent-kaurene synthase—has been identified that catalyzes the conversion of geranyl-geranyl pyrophosphate to ent-kaurene, which in turn can be oxidized and rearranged to give GA_{12} -aldehyde ***(11)***. For cytokinins a cell-free system has been isolated from the slime mold *Dictostelium discordeum*, which can use adenosine monophosphate (AMP), and the isopentenyl group from isoprenyl pyrophoshate, to synthesize *trans*-zeatin.

The final major plant hormone is abscisic acid (ABA), and again much is known concerning the biosynthesis of this compound. ABA is a sesqueterpenoid with close structural similarities to the terminal rings of several oxygenated carotenoids (xanthophylls), such as violaxanthin. An ABA-deficient mutant of *Arabidopsis thaliana* was shown to be impaired in the epoxidation of zeaxanthin to form violaxanthin ***(12,13)***. This gene was subsequently cloned and a tomato homolog has recently been obtained ***(14)***. This provides strong evidence that ABA is biosynthesised from xanthophyll precursors. Further evidence has been provided by a transposon-induced ABA-deficient mutant of maize (the *vp-14* mutant allele). This gene has been cloned and a fusion protein expressed in *E. coli* ***(15)***. The enzyme was shown to carry out the oxidative cleavage of 9-cis xanthophylls to form the first C_{15} intermediate in the ABA pathway, an aldehyde known as xanthoxin. A tomato homolog of this has recently been obtained and it has been shown that mRNA levels are upregulated by water stress ***(16)***.

Cell-free systems can be produced that carry out the conversion of xanthoxin to ABA ***(17,18)***. The penultimate precursor of ABA has been shown to be ABA-aldehyde (AB-CHO) ***(18,19)*** and the enzyme carrying out this conversion can thus be called ABA-aldehyde oxidase (AAO) ***(20)***. The gene encoding this enzyme has yet to be cloned.

In this chapter the extraction and purification of AAO will be used as an example to illustrate the general principles involved in the isolation of key enzymes involved in hormone synthesis. The extraction described is based on that of Sindhu and Walton ***(18)***. Although precise methods will be given for the extraction and assay of AAO it is obvious that these will not be directly applicable to other hormone systems. Indeed, it is very difficult to predict the conditions required for the extraction and assay of any enzyme. The purification methods described, apart from the affinity chromatography, are obviously generic and can be applied to any enzyme system. Again, however, the conditions required would need to be determined by experiment for each individual case. Some general considerations are given in **Notes 1–6**. Several excellent texts cover in detail the considerations required in protein purification ***(21–24)***. The authors hope that this chapter will provide an idea of the general approach to be taken.

2. Materials

2.1. Extraction and Assay

1. Plant tissue: Leaf tissue used in these experiments was obtained from tomato plants (*Lycopersicum esculentum*, Mill cv. Ailsa Craig). The experimental material was grown in a glasshouse. Supplementary lighting was supplied to give a light period of at least 16 h and a dark period of no less than 8 h. The age at which the plants were harvested was determined by the number of true leaves. Plants having between four and 10 true leaves were harvested. In any one extraction there was a range of tissue ages.
2. Extraction and assay buffer: 0.2 *M* KH_2PO_4 , 7.5 m*M* dithiothreitol adjusted to pH 6.5 with 0.1 *M* KOH.
3. Substrate: 0.1 mg/mL^{-1} *c*-(*RS*)-AB-CHO in acetone.
4. Partitioning of ABA: Diethyl ether. Rotary evaporator, Howe Gyrovap (Banbury, UK).
5. High performance liquid chromatography (HPLC) analysis of ABA.
   a. Equipment: Perkin-Elmer (Beaconsfield, UK) series 10 Liquid Chromatograph, Perkin-Elmer LC-95 UV/Visible Spectrophotometer Detector, and Perkin-Elmer LCI-100 Laboratory Computing Integrator.
   b. Column: Analytical reverse-phase either a 10 µm Techsil C_{18} ODS bonded column (250 × 4.6 mm id) or a Bio-Rad (Hemel Hempstead, UK) Rsil C_{18} ODS bonded column (250 × 4.6 mm id).
   c. Eluent: 45% aqueous methanol (Romil, UK) acidified with 1% ethanoic acid de-gassed *in vacuo*.
6. GC-MS analysis of ABA:
   a. Equipment: Hewlett Packard (Palo Alto, CA) 5890A series gas chromatograph linked to a Hewlett Packard 5970 series mass selective detector. Grob type splitless injector (1 µL sample volume). BP-1 WCOT 25 m × 0.22 mm id column.

b. Diazomethane reagent: React 10.7 g diazald (Aldrich Chemical Company Ltd., UK) in 65.mL diethyl ether with 5 g potassium hydroxide (KOH) dissolved in 33 mL of 68% aqueous ethanol at 40°C. Collect the resultant diazomethane in diethyl ether at 0°C.

2.2. Affinity Chromatography

1. Sepharose 6B support with an epoxy-activated 12-atom spacer (Sigma Chemical Company Ltd, UK).
2. (*RS*)-*cis*-ABA.
3. Column buffer: Eluent from ultrafiltration buffered with 50 m*M* Tris-HCl, pH 7.5.
4. 42 μ*M* AB-CHO in column buffer.

2.3. Protein Determination

1. Bio-Rad protein determination kit.
2. Bovine serum albumin (Sigma).
3. Spectrophotometer.

2.4. SDS Gel Electrophoresis

2.4.1. Preparing and Running the Gel

1. 30% Acrylamide, 0.8% *NN'*-methylenebisacrylamide (BDH, UK) made up in distilled H_2O and filtered through a Whatman No. 1 filter paper.
2. Tris-HCl, pH 8.8 (at 20°C).
3. 1% SDS (sodium dodecyl sulfate) (Fisons, UK).
4. *N,N,N',N'*-Tetramethylethylanadiamine (TEMED) (Sigma).
5. Freshly prepared 5% (w/v) aqueous ammonium persulfate.
6. 1*M* Tris-HCl, pH 6.8 (at 20°C).
7. Electrode buffer: 0.192 *M* glycine, 0.025 *M* Tris base, 0.1% (w/v) SDS pH 8.8. This was prepared fresh for each run.
8. Sample buffer: 10% (w/v) SDS, 20% (v/v) glycerol (Fisons, UK), 0.04% (w/v) bromophenol blue (Fisons), 125 m*M* Tris-HCl, pH 6.8, and 100 m*M* iodoacetamide (Sigma).
9. Pharmacia minigel system (Pharmacia, Sweden). Gallekamp Biomed E250 power pack (Fisons, UK).

2.4.2. Staining Gel with Coomassie Blue

1. 45% (v/v) Methanol (Fisons), 10% (v/v) ethanoic acid, and 0.1% (w/v) Coomassie Brilliant Blue R-250 (Fisons).
2. 45% (v/v) Methanol, 10% (v/v) ethanoic acid
3. 20% (v/v) Methanol, 10% (v/v) ethanoic acid

2.4.3. Silver Staining the Gel

1. 50% Ethanol (Fisons); 10% ethanoic acid (Fisons).
2. 25% Ethanol; 10% ethanoic acid.
3. 10% Ethanol; 0.5% ethanoic acid.
4. 11 m*M* silver nitrate ($AgNO_3$) (Fisons).
5. 95 m*M* Methanol (Fisons); 750 m*M* NaOH (Fisons).
6. 10% Ethanoic acid.

3. Methods

3.1. Extraction and Assay

1. Leaf tissue (typically 5–20 g), including cotyledons, was harvested and weighed. The tissue was blended in a blender (Moulinex, France) using 3 mL cold extraction buffer per gram fresh weight of leaf tissue. The homogenate was filtered through four layers of muslin and centrifuged at 20,000*g* for 30 min at 4°C. The supernatant was retained and the pellet discarded.
2. Crude extract was concentrated using a stirred cell ultrafiltration unit (Amicon, UK), fitted with a filtration membrane with a mol wt cut-off of 1000 kDa (PM10 membrane, Amicon, UK). The sample was filtrated under nitrogen at a pressure of 60 psi. Ultrafiltration was carried out over ice. Both the eluent and retentate were retained.
3. Crude extract (0.3–0.5 mL) was made up to a total volume of 0.7 mL with assay buffer and 0.75 mL of eluent from the ultrafiltration added. The requirement for eluent is discussed in **Note 4**.
4. Fifty microliters of substrate was added to give a final substrate concentration of 8.06 μ*M*. The assay mixture was briefly mixed (1 s) using a Whirlimixer and to determine *c*-ABA concentration at time = 0, a 0.5-mL aliquot was taken and frozen immediately in liquid nitrogen.

5. The remaining 1 mL of assay mixture was incubated for 60 min at 32°C, then frozen in liquid nitrogen. The samples were stored at –20°C until extraction of ABA for analysis and quantification.
6. The samples were defrosted and the pH was adjusted to pH 3.0 by the addition of 10% aqueous HCl. The samples were then partitioned against a fivefold excess of diethyl ether (Fisons Ltd., UK, redistilled, or Romil Ltd., UK). The aqueous phase was discarded.
7. The organic phase was taken to dryness at 30°C *in vacuo*, using a rotary evaporator.
8. The samples were redissolved in 500 µL methanol (Romil Ltd., UK) and transferred to a 900 µL microinsert (Phase Sep, UK). The samples were taken to dryness at 20°C *in vacuo* using a Howe Gyrovap. The samples were stored at 4°C until analysis by HPLC.
9. The samples were redissolved in 50 µL methanol. A 20-µL aliquot was injected onto the HPLC column, using a fixed volume sample loop, via a Rheodyne 7125 valve. Analytical reverse-phase HPLC was carried out on either a 10 µm Techsil C_{18} ODS bonded column (250 × 4.6 mm id) or a Bio-Rad Rsil C_{18} ODS bonded column (250 × 4.6 mm id). The column conditions were 45% aqueous methanol (Romil, UK) acidified with 1% ethanoic acid. The eluent was degassed *in vacuo*. An isocratic system was used. The solvent flow rate was 2 mL/min, delivered by a Perkin-Elmer series 10 Liquid Chromatograph. Eluted compounds were detected at 260 nm using a Perkin-Elmer LC-95 UV/Visible Spectrophotometer Detector, and the peak area and peak height calculated by a Perkin-Elmer LCI-100 Laboratory Computing Integrator.
10. Confirmation of the identity of the *cis*-ABA peak was carried out by comparison of retention time with the peak obtained from an authentic external standard. In addition, validation of the peak identity was carried out by both gas chromatography-mass spectroscopy (GC-MS) and ultraviolet (UV) scan.
11. Confirmation by GC-MS was carried out by taking the peak from the HPLC, which eluted at the retention time of authentic *c*-ABA standards. This was taken to dryness at 35°C *in vacuo* using a rotary evaporator. The sample was redissolved in methanol and transferred to a 900-µL microinsert. The sample was then taken to dryness at 20°C *in vacuo* using a Howe Gyrovap.
12. The sample was derivatized by methylation (to give 1-methyl-ABA ester (Me-ABA). One milliliter of diazomethane solution was added

to the sample. The sample was left at room temperature (in a fume cupboard) and allowed to react for 2 h. The sample was dried at 20°C *in vacuo* using a Howe Gyrovap, and redissolved in 20 µL of 10 µg/mL Et-ABA for quantification by GC-MS.

13. GC-MS was performed on a Hewlett Packard 5890A series gas chromatograph linked to a Hewlett Packard 5970 series mass selective detector. Sample aliquots of 1 µL were injected via a Grob type splitless injector onto a BP-1 WCOT 25 m × 0.22 mm id column. The carrier gas was helium, at a flow rate of 1 mL/min. The temperature program started at 135°C and was maintained for 2 min before being increased to a final temperature of 265°C at a rate of 10°C/min. Mass spectra for eluted compounds were recorded over the range m/z 40–320; scan time was 0.5 s. Confirmation of the identity of the *cis*-ABA was by comparison of mass spectra obtained from authentic standards and published data ***(25)***.

3.2. Purification

Many methods are available for the purification of proteins and no generic method can be described. A very powerful and specific purification method, however, is affinity chromatography. It is not always applicable but given the availability of a suitable ligand often provides a useful single-step purification. The construction and use of an affinity column for the purification of AAO will be described here.

1. Both *cis*- and *trans*-diol isomers of 1',4'-ABA-diol (1',4'-ABA-*cis/trans*-diol) were used as the ligand. This was synthesized from (*RS*)-*cis*-ABA (Lancaster Synthesis, UK) by reduction of the 4'-keto group by sodium borohydride, using the method described by Milborrow ***(26)***: 0.775 g (2.93 mmol) (*RS*)-*c*-ABA was dissolved in 16 mL methanol. To this 2 g sodium borohydride was added, dissolved in 8 mL H_2O. The reaction mixture was kept on ice for 30 min, and the reaction stopped by the addition of 95% ethanoic acid. The methanol was evaporated off at 30°C *in vacuo*. The aqueous phase was partitioned against a twofold excess of diethyl ether. The organic phase was retained and taken to dryness at 30°C *in vacuo*. All reactions were carried out under dimmed lighting conditions to prevent isomerization of the 2,3-C double bond.

2. The matrix used was a Sepharose 6B support with an epoxy-activated 12-atom spacer (Sigma Chemical Company Ltd, UK). Five grams of epoxy-activated Sepharose 6B (285-600 μmol activated sites) were hydrated overnight in 20 m*M* Na_2HPO_4, adjusted to pH 10.75 using 2 *M* NaOH. The matrix was washed with 100 mL of the hydration buffer and allowed to drain to remove excess buffer.
3. For coupling the ligand to the matrix, the 1',4'-ABA-diol was redissolved in 2 mL acetone. To this was added 58 mL of 20 m*M* Na_2HPO_4, adjusted to pH 10.75 using 2 *M* NaOH. The hydrated matrix was added to the mixture, and the coupling reaction was allowed to proceed for 16 h at 35°C, with continuous shaking.
4. The slurry was transferred to a GF/A filter and washed with 300 mL dH_2O, 500 mL 100 m*M* sodium carbonate-bicarbonate buffer, pH 8.0, and 500 mL of 100 m*M* sodium acetate-ethanoic acid buffer, pH 4.
5. To block any nonreacted sites, the gel was then reacted with 1 *M* ethanolamine for 4 h at room temperature, at pH 9.0, using a Na_2HPO_4 buffer adjusted to pH 9.0 using 2 *M* NaOH. The slurry was then washed as above before being equilibrated with 200 m*M* Tris-HCl, pH 7.5, and packed into a column. The affinity gel matrix was stored at 4°C.
6. The affinity column was equilibrated with a column buffer (Eluent from ultrafiltration buffered with 50 m*M* Tris-Cl, pH 7.5, *see* **Note 7**). One milliliter of extract was applied and washed through with column buffer. ABA oxidase activity was eluted by the application of 42 μ*M* AB-CHO in column buffer, *see* **Note 8**. Fractions were collected and assayed for enzyme activity and protein content.
7. A typical elution profile from the affinity column is shown in Fig. 1. The specific activity of the crude extract was approx 4 nmol ABA/hr/mg protein and of the affinity purified enzyme about 230 nmol ABA/hr/mg protein.

3.3. Protein Determination

Protein determination was carried out using a Bio-Rad protein assay microassay procedure, *see* **Note 9**.

1. Protein samples were diluted with 20 m*M* Tris-HCl buffer, pH 7.5, to give a concentration within the range of 1–20 μg/mL.
2. E_{595} was measured on a Pye Unicam SP8-400 UV/visible spectrophotometer.

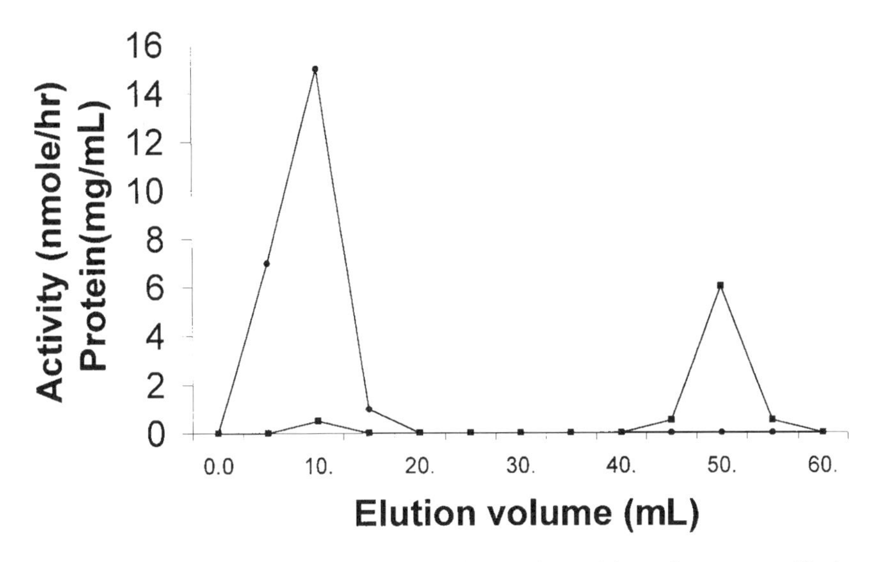

Fig. 1. Ellution profile of ABA-aldehyde oxidase from an affinity column. (●) Protein. (■) Enzyme activity.

3. Standard calibration curves were constructed using a concentration range of bovine serum albumin (Sigma) of 0–25 μg/mL.

3.4. SDS Gel Electrophoresis

Denaturing, discontinuous sodium dodecyl sulfate-polyacrylamide gel electrophoresis (SDS-PAGE) was used for analysis of protein purity and for estimation of M_r of subunits. SDS-PAGE was performed with a Pharmacia minigel system.

3.4.1. Preparing and Running the Gel

1. For the main gel and gel plug, a solution (solution A) of the following constituents was made up: 3.67mL 30% acrylamide/0.8% *NN'*-methylenebisacrylamide, 4.18 mL Tris-HCl, pH 8.8 (at 20°C), 1.1 mL 1% SDS (sodium dodecyl sulfate), and 2.05 mL H_2O (distilled). The solution was de-gassed in vacuo.
2. Two milliliters of solution A were taken. To this was added: 4 μL *N,N,N',N'*-tetramethylethylanadiamine (TEMED) and 0.1 mL freshly prepared 5% (w/v) aqueous ammonium persulfate. This was poured

into the gel mold and allowed to set to give a plug that acted as a seal at the bottom of the gel.

3. To the remainder of solution A was added: 4.4 µL TEMED and 0.2 mL freshly prepared 5% (w/v) aqueous ammonium persulfate. The gel was then poured. The exposed surface of the gel was covered with a layer of water after pouring, to prevent dehydration and to give a flat surface. The gel was allowed to set for at least 45 min.
4. For the stacking gel, a solution of the following solution was made up: 0.66 mL 30% acrylamide/0.8% *NN'*-methylenebisacrylamide, 0.5 mL 1 *M* Tris-HCl, pH 6.8 (at 20°C), 1 mL 1% SDS, 6.04 mL H_2O (distilled), 2 µL TEMED, and 0.1 mL freshly prepared 5% (w/v) aqueous ammonium persulfate.
5. The water was removed from the surface of the set main gel and replaced with stacking gel solution. A suitable gel comb was inserted and the stacking gel allowed to set. Prior to sample loading the comb was removed and sample wells were flushed with 50 µL of electrode buffer to remove any nonpolymerized polyacrylamide. The electrode tanks were then filled with electrode buffer.
6. The protein samples for SDS-PAGE were added to an equal volume of sample buffer. The protein samples were then heated for 5 min in a boiling water bath. The samples were spun at 8400*g* for 5 min to precipitate any particulate matter resulting from the heat treatment.
7. Protein samples (20–50 µL) were loaded into the wells. The gel was run at 30 mV until the blue tracker dye had passed into the plug.

3.4.2. Staining the Gel

3.4.2.1. Coomassie Blue Staining

1. The following aqueous solutions were made up:
   A. 45% (v/v) Methanol (Fisons), 10% (v/v) ethanoic acid, and 0.1% (w/v) Coomassie Brilliant Blue R-250.
   B. 45% (v/v) Methanol and 10% (v/v) ethanoic acid.
   C. 20% (v/v) Methanol and 10% (v/v) ethanoic acid.
2. Fixing and staining of the proteins after electrophoresis was accomplished by immersing the gel in solution A for 4 h.
3. Destaining (removal of background stain) was carried out by immersing the gel in solution B for 24 h. Two changes of solution B were used to enhance the destaining process.
4. The gel was stored in solution C.

3.4.2.2. Silver Staining

1. The following aqueous solutions were made up:
   A. 50% Ethanol and 10% ethanoic acid.
   B. 25% Ethanol 10% ethanoic acid.
   C. 10% Ethanol and 0.5% ethanoic acid.
   D. 11 m*M* silver nitrate.
   E. 95 m*M* Methanol and 750 m*M* NaOH.
   F. 10% Ethanoic acid.
2. Proteins were fixed in the gel by immersing the gel in solution A for 1 h.
3. Rehydration of the gel and partial removal of ethanoic acid was achieved by immersing the gel in solution B, followed by solution C, both for 30 min.
4. The gel was immersed in solution D for an additional 1 h.
5. The gel was then rinsed rapidly (2 × 45 s) in distilled water to remove unbound $AgNO_3$ from the gel surface.
6. Color development was carried out by immersing the gel in solution E until bands were visible.
7. The reaction was stopped by immersion of the gel in solution F.
8. The gels were stored at 4°C in solution B.
9. *See* **Notes 10** and **11**.

4. Notes

1. During the extraction, enzyme stability can be influenced by many parameters. These need to be determined experimentally for each specific enzyme. However, there are some general precautions. The ionic strength within a cell is typically within the range of 0.15–0.2 *M*. It is thus normal to use this ionic strength to solubilize cytoplasmic proteins. However, the cell wall can act as a strong ion-exchange medium and many basic proteins can adhere to the wall remnants following homogenization. These can often be desorbed using higher ionic strength (1 *M*) and neutral pH. Similarly hydrophobic proteins, and those proteins that are intrinsically membrane associated, can often be solubilized by the addition of a nonionic detergent, such as Triton X-100 (0.1%–3%) or by the use of chaotropic agents, such as 1 *M* LiCl. It should be noted, however, that these reagents can themselves result in enzyme inactivation.
2. The extraction solution should also be buffered. Typical buffers include 50 m*M* phosphate and 0.1 *M* Tris-HCl. The normal pH range

within a cell is between 6.5 and 7.5, and this pH range is a good starting point for any extraction. It should be noted that the pH of many buffer systems, in particular Tris, will vary with temperature. Thus, storage of extracted enzyme at low temperature can sometimes induce adverse pH changes, resulting in inactivation of the enzyme.

3. The active site of an enzyme can be modified by many factors, including oxidation and heavy metals. Oxidation is normally prevented by the addition of reducing agents, those commonly employed being β-mercaptoethanol (5%), cysteine (10 m*M*), or dithiothreitol (DTT) (5–10 m*M*). However, all three of these reducing agents will oxidize rapidly in solution and should always be used fresh. Of the three, DTT is perhaps preferred since this does not form mixed disulfides with proteins. It is often necessary to include a chelating agent in the extraction buffer. Ethylenediamine tetra-acetic acid (EDTA) at between 1 and 5 m*M* is commonly used.
4. For many enzymes, essential cofactors are often required both for activity and to retain stability during purification. These may include metal ions, such as copper and iron. For instance, ACC oxidase requires Fe^{2+} and ascorbate. The requirement of ABA oxidase for eluent from the initial ultrafiltration of the crude extract probably indicates the presence of an essential cofactor. This has not been identified but could be molybdenum since it is postulated that ABA oxidase contains a molybdenum cofactor.
5. During extraction it is often necessary to protect the enzyme from proteolytic activity. A range of protease inhibitors can be employed, including phenylmethylsulfonyl fluoride (PMSF). In addition, the inclusion of a chelating agent may also serve to limit some forms of protease.
6. Dilute enzyme solutions tend to lose activity more rapidly than concentrated solutions. It is thus advisable to concentrate the samples following each purification step as quickly as possible. Several methods are available for concentrating proteins. These include ultrafiltration, salting out with ammonium sulfate, and precipitation with organic solvents, such as ethanol. It is also a common practice to stabilize very dilute proteins by the addition of bovine serum albumin (0.1–10 mg/mL) or glycerol (20–30%). Glycerol can be used at up to 50% but under these conditions becomes too viscous to handle. However, it is common to use 50% glycerol in solution to be stored at low temperature and under these conditions the effect on the retention of activity can be very dramatic.

7. The affinity column gave best recovery when the column buffer was made using the eluent from the initial ultrafiltration step. The reason for this is unclear but may indicate the presence of a limiting cofactor. The nature of this cofactor was not determined. However, since it has been suggested *(27)* that ABA oxidase may require a molybdenum cofactor, this may be involved.
8. Ionic strength of the column buffer may influence the efficiency of both binding and elution from the affinity column. ABA, rather than AB-CHO, was equally effective at eluting the enzyme from the column. However, in this case the residual ABA in the fractions can interfere with subsequent assay. Fractions in this case would thus need desalting using spun G-25 columns prior to the assay.
9. Several methods are available for protein determination. The Bio-Rad microassay is sensitive and, unlike some of the other methods, does not appear to be unduly effected by the reagents commonly used for enzyme extraction, such as Tris, DTT, and ammonium salts. However, it has been noticed with several plant enzymes that the absolute protein content as determined with the Bio-Rad system is often much lower than that determined by the Lowry method. Thus, it is important to realize that these assays can often only provide a relative measure of protein content and are not absolute.
10. Coomassie staining is linear up to around 20 µg/cm but intensity varies between individual proteins. Silver staining is 10–20 times more sensitive than Coomassie staining but this again depends on the type of protein. In a crude extract it is normal to load between 50 and 100 µg of protein for Coomassie staining and between 10 and 25 µg for silver staining.
11. The silver staining method is also very prone to the appearance of false bands. The reason for this is unclear but may be caused by keratin or the gel chemistry. The incorporation of iodoacetamide in the sample loading solution minimizes this problem.

References

1. Adams, D. O. and Yang, S. F. (1979) Ethylene biosynthesis: identification of 1–aminocyclopropane-1–carboxylic acid as an intermediate in the conversion of methionine to ethylene. *Proc. Natl. Acad. Sci. USA* **76,** 170–174.

2. Hamilton, A. J., Lycett, G. W., and Grierson, D. (1990) Antisense gene that inhibits synthesis of the hormone ethylene in transgenic plants. *Nature* **346,** 284–287.
3. Hamilton, A. J., Bouzayen, M., and Grierson, D. (1991) Identification of a tomato gene for the ethylene forming enzyme by expression in yeast. *Proc. Natl. Acad. Sci. USA* **88,** 7434–7437.
4. Ververdis, P. and John, P. (1991) Complete recovery in vitro of ethylene forming enzyme activity. *Phytochemistry* **30,** 725–727.
5. Sato, T. and Theologis, A. (1989) Cloning the mRNA encoding 1-aminocyclopropane-1-carboylate synthase, the key enzyme for ethylene biosynthesis in plants. *Proc. Natl. Acad. Sci. USA* **86,** 6621–6625.
6. Oeller, P. W., Min-Wong, L., Taylor, L. P., Pike, D. A., and Theologis, A. (1991) Reversible inhibition of tomato fruit senescence by antisense RNA. *Science* **254,** 437–439.
7. MacMillan, J. (ed.) (1980) Hormonal regulation of development, in *Encyclopedia of Plant Physiology*, vol. 9, Springer-Verlag, Berlin.
8. Crozier, A. and Hillman, J. R. (1984) The biosynthesis and metabolism of plant hormones, in *Society for Experimental Biology Seminar Series*, no 23, Cambridge University Press, Cambridge, UK.
9. Bartel, B. (1997) Auxin biosynthesis. *Ann. Rev. Plant Physiol. Plant Mol. Biol.* **48,** 51–66.
10. Binns, A. N. (1994). Cytokinin accumulation and action: Biochemical, genetic and molecular approaches. *Ann. Rev. Plant Physiol. Plant Mol. Biol.* **45,** 173–196.
11. Hedden, P. and Kamiya, Y. (1997) Gibberellin biosynthesis: enzymes, genes and their regulation. *Ann. Rev. Plant Physiol. Plant Mol. Biol.* **48,** 431–460.
12. Duckham, S. C., Linforth, R. S. T., and Taylor, I. (1991) Abscisic acid deficient mutants at the aba gene locus of Arabidopsis thaliana are impaired in the epoxidation of zeaxanthin. *Plant Cell Environ.* **14,** 601–606.
13. Rock, C. D. and Zeevaart, J. A. D. (1991) The aba mutant of Arabidopsis thaliana is impaired in epoxy-carotenoid biosynthesis. *Proc. Natl. Acad. Sci. USA* **88,** 7496–7499.
14. Burbidge, A., Grieve,T., Terry, C., Corbett, J., Thompson, A., and Taylor, I. (1997) Structure and expression of a cDNA encoding zeaxanthin epoxidase, isolated from a wilt-related tomato (*Lycopersicon esculentum* Mill) library. *J. Exp. Bot.* **48,** 1749,1750.

15. Schwartz, S. H., Tan, B. C., Gaga, D. A., Zeevaart, J. A. D., and McCarthy, D. R. (1997) Specific oxidative cleavage of carotenoids by VP14 of maize. *Science* **276,** 1872–1874.
16. Burbidge, A., Grieve, T., Jackson, A., Thompson, A., and Taylor, I. (1997) Structure and expression of a cDNA encoding a putative neoxanthin cleavage enzyme (NCE), isolated from a wilt-related tomato (*Lycopersicon esculentum* Mill) library. *J. Exp. Bot.* **48,** 2111,2112.
17. Sindhu, R. K. and Walton, D. C. (1987) Conversion of xanthoxin to abscisic acid by cell-free preparations from bean leaves. *Plant Physiol.* **85,** 916–921.
18. Sindhu, R. K. and Walton, D. C. (1988) Xanthoxin metabolism in cell-free preparations from wild-type and wilty tomato mutants. *Plant Physiol.* **88,** 178–182.
19. Taylor, I. B., Linforth, R. S. T., Al-Naieb, R. J., Bowman, W. R., and Marples, B. A. (1988) The wilty tomato mutants *flacca* and *sitiens* are impaired in the oxidation of ABA-aldehyde to ABA. *Plant Cell Environ.* **11,** 739–745.
20. Sindhu, R. K., Griffin, D. H., and Walton, D. C. (1990) Abscisic alhdehyde is an intermediate in the enzymic conversion of xanthoxin to abscisic acid in *Phaseolus vulgaris L.* leaves. *Plant Physiol.* **93,** 689–694.
21. Deutscher, M. P. (1990) Guide to protein purification. *Methods Enzymol.* **182.**
22. Harris, E. L. V. and Anagal, S. (1990) *Protein Purification Methods: A Practical Approach.* IRL, Oxford, UK.
23. Harris, E. L. V. and Anagal, S. (1990) *Protein Purification Applications: A Practical Approach.* IRL, Oxford, UK.
24. Scopes, K. S (1993) *Protein Purification: Principles and Practice.* Springer-Verlag, New York.
25. Netting, A. G., Milborrow, B. V., and Duffield, A. M. (1982) Determination of abscisic acid in *Eucalyptus haemastoma* leaves using gas chromatography/mass spectrometry and deuterated internal standards. *Phytochemistry* **21,** 385–389.
26. Milborrow, B. V. (1984) The conformation of abscisic acid by n. m. r and a revision of the proposed mechanism for cyclization during biosynthesis. *Biochem. J.* **220,** 325–332.
27. Walker-Simmons, M., Kudrna, D. A., and Warner, R. L. (1989) Reduced accumulation of ABA during water stress in a molybdenum cofactor mutant of barley. *Plant Physiol.* **90,** 728–733.

2

Differential Display

Analysis of Gene Expression During Plant Development

Catherine A. Whitelaw, Benedetto Ruperti, and Jeremy A. Roberts

1. Introduction

Differential display has been developed as a tool to detect and characterize altered gene expression in eukaryotic cells. In this technique, as first described by Liang and Pardee (1992) *(1,2)* total RNA is isolated from two (or more) cell types or tissues to be compared. First strand copies of both RNAs are made by reverse transcription, using an oligo-dT primer (the anchor primer) that has a specific dinucleotide at its 3' end (*see* **Table 1**). This anchor primer and a random 10-mer (arbitrary primer) (**Table 1**) are then used to amplify, by polymerase chain reaction (PCR), cDNAs to which the 3'-anchor and 5'-arbitrary primers both hybridize (**Fig. 1**). A radioactive nucleotide is included in the PCR reactions so that the PCR products can be run side-by-side on a 5% polyacrylamide gel and visualized by autoradiography.

Bands that appear on the display of RNA from one cell type but not the other correspond to differentially expressed mRNAs. These bands are excised from the gel and reamplified with the same primers used for the original display. The resulting PCR product can

From: *Methods in Molecular Biology, vol. 141: Plant Hormone Protocols*
Edited by: G. A. Tucker and J. A. Roberts © Humana Press Inc., Totowa, NJ

Table 1
Oligonucleotide Primers Used in Differential Display

	Anchor primers		Arbitrary primers
1	5'- TTT TTT TTT TTT (AGC)A -3'	A	5'- AGC CAG CGA A -3'
2	5'- TTT TTT TTT TTT (AGC)C -3'	B	5'- GAC CGC TTG T -3'
3	5'- TTT TTT TTT TTT (AGC)G -3'	C	5'- AGG TGA CCG T -3'
4	5'- TTT TTT TTT TTT (AGC)T -3'	D	5'- GGT ACT CCA C -3'
5	5'- TTT TTT TTT TTT GC -3'	E	5'- GTT GCG ATC A -3'[a]
6	5'- TTT TTT TTT TTT GT -3'		
7	5'- TTT TTT TTT TTT GG -3'		
8	5'- TTT TTT TTT TTT GA -3'		
9	5'- TTT TTT TTT TTT CC -3'		
10	5'- TTT TTT TTT TTT CT -3'		
11	5'- TTT TTT TTT TTT CG -3'		
12	5'- TTT TTT TTT TTT CA -3'		

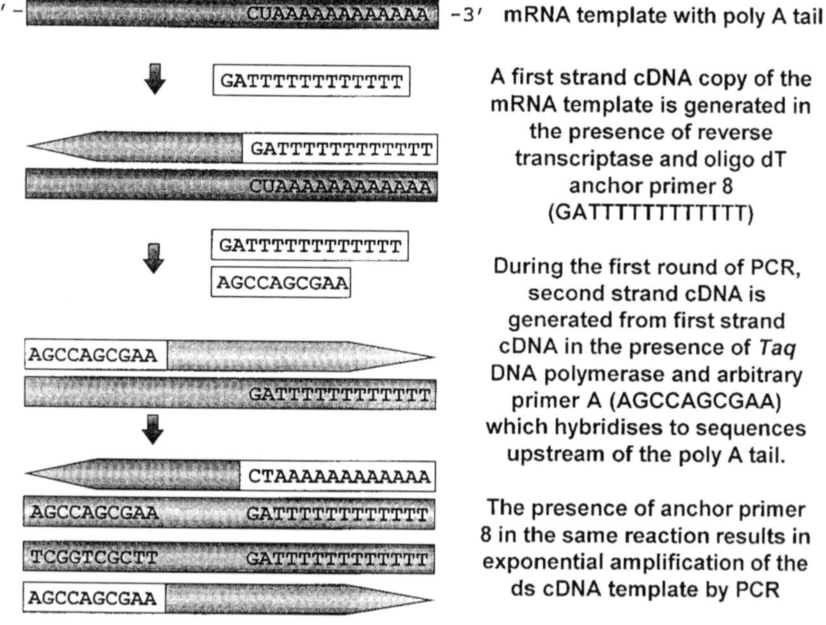

Fig. 1. Diagram outlining the major enzymic reactions involved in differential display.

then be cloned and sequenced or used directly to probe Northern blots. Once confirmation of differential expression is obtained, the PCR products can be used to isolate full length cDNAs from appropriate cDNA libraries.

We have applied the differential display technique to compare changes in gene expression during plant developmental processes. In particular, we are interested in isolating genes involved in cell separation events, culminating in leaf abscission and pod dehiscence in oilseed rape (Brassica napus). Both processes involve coordinated cell wall breakdown and cell separation at specific sites, accompanied by increases in the activity of several cell wall-degrading enzymes. Ethylene is known to promote abscission but the role of this gaseous plant hormone in dehiscence is less clearly defined. Differential display has proven to be a successful technique, resulting in the isolation of a number of rare and novel abscission- and dehiscence-related transcripts.

2. Materials

Unless otherwise stated, all chemicals used in this protocol were analytical grade. The researcher is advised to consult the appropriate control of substances hazardous to health (COSHH) regulations wherever stated.

2.1. DNase Treatment of RNA

1. RQ1 RNase-free DNase (1 U/µL) (Promega [Madison, WI]).
2. RNasin® ribonuclease inhibitor (20–40U/µL) (Promega).
3. Tris-buffered phenol, pH 7.9 (ICN [Costa Mesa, CA]) (COSHH).
4. Chloroform (BDH [Dorset, England]) (COSHH).
5. 3 *M* sodium acetate, pH 6.0 (RNase-free).
6. 100% Ethanol.
7. 70% (v/v) Ethanol.

2.2. First Strand cDNA Synthesis

1. Thermocycler (or water baths set at 65°, 37°, and 95°C).
2. Anchor primer (200 µ*M*) (*see* **Table 1**).
3. Moloney murine leukemia virus (M-MLV) reverse transcriptase (50 U/µL) (Stratagene [La Jolla, CA]). This is supplied with 5X reaction buffer.

4. Ultrapure dNTPs (100 m*M* solutions) (Pharmacia [Piscataway, NJ]).
5. RNasin® ribonuclease inhibitor (20–40 U/μL) (Promega).

2.3. PCR Amplification

1. Thermocycler.
2. Ultrapure dNTPs (Pharmacia).
3. *Taq* DNA Polymerase (5 U/μL) (Gibco-BRL [Rockville, MD]). This is supplied with 10X buffer and 50 m*M* $MgCl_2$.
4. Anchor primer (200 μ*M*); the same anchor primer that was used in the first-strand cDNA synthesis reaction.
5. Arbitrary primer (50 μ*M*) (*see* **Table 1**).
6. ^{35}S-dATP (Amersham AG1000 > 1000 Ci/mmol [Pharmacia]). This nucleotide is radioactive, therefore local rules for use, storage, and disposal should be followed.
7. Mineral oil (Sigma [St. Louis, MO] molecular biology grade).

2.4. Denaturing Polyacrylamide Gel Electrophoresis and Autoradiography

1. Sequencing apparatus, including glass plates, clamps, spacers, and sharkstooth comb (Bio-Rad).
2. Power pack (capable of supplying 2000 V).
3. Gel dryer and vacuum pump (Bio-Rad).
4. Masking tape.
5. Gel Slick™ (AT Biochem [Cambridge, England]).
6. 20 mL syringe.
7. Whatman 3MM filter paper.
8. Saran Wrap (or cling film).
9. X-ray film (BioMax-MR, Kodak, supplied by Anachem [Luton, England]).
10. Long Ranger™ gel solution (50X concentrate) (FMC Bioproducts [Rockland, ME) (COSHH).
11. TEMED (*N,N,N',N'*-tetramethylethylenediamine) (Sigma) (COSHH).
12. Ammonium persulphate (APS) (Sigma) (COSHH). A 10% (w/v) solution is prepared by adding 0.2 g APS to 2 mL dH_2O. A 25% (w/v) solution is prepared by adding 0.5 g APS to 2 mL dH_2O. These solutions can be stored at 4°C for 1 wk.
13. Urea (Gibco-BRL).

14. 10X TBE running buffer. For 1 L, add 108 g Tris base (BDH [Dorset, England]), 55 g boric acid (BDH), and 9.3 g ethylenediamine tetraacetic acid (EDTA) to 800 mL dH_2O. Allow to dissolve and then make the volume up to 1 L with distilled water (dH_2O) (*see* **Note 1**). The working concentration of TBE is 1X.
15. Gel-loading buffer (95% [v/v] formamide [Sigma] [COSHH], 20 m*M* EDTA [BDH], 0.05% [w/v] xylene cyanol [Sigma], 0.05% [w/v] bromophenol blue [Sigma]).
16. Water bath set at 80°C.

2.5. Elution and Reamplification of DNA Fragments

1. Glycogen (Sigma) 10 mg/mL solution in dH_2O.
2. 3 *M* sodium acetate, pH 5.2 (autoclaved).
3. 100% Ethanol.
4. 85% (v/v) Ethanol.
5. Thermocycler.
6. Ultrapure dNTPs (Pharmacia).
7. *Taq* DNA polymerase (5 U/μL) (Gibco-BRL). This is supplied with 10X reaction buffer and 50 m*M* $MgCl_2$.
8. Anchor primer (200 μ*M*); the same anchor primer that was used in the differential display PCR reaction.
9. Arbitrary primer (50 μM); the same arbitrary primer that was used in the differential display PCR reaction.
10. Mineral oil (Sigma, molecular biology grade).
11. Low-melting-point agarose (NuSieve GTG, FMC Bioproducts).

3. Methods

3.1. DNase Treatment

The RNA is treated with RNAse-free DNAse in order to remove any contaminating genomic DNA (*see* **Note 2**).

1. Add the following components to a RNAse-free microfuge tube:

Component	Amount required	Final concentration
RNA	10 μg	10 μg
DNase I (1 U/μL)	10 μL	10 U
RNase inhibitor (20–40 U/μL)	1 μL	20–40 U
dH_2O	to 100 μL	

2. Vortex and centrifuge briefly to bring the contents to the bottom of the tube.
3. Incubate the reaction at 37°C for 30 min.
4. Add 100 µL Tris-buffered phenol, pH 7.9, vortex for 15 s, then spin for 5 min at maximum speed (~13,000 rpm) in a bench-top microcentrifuge.
5. Transfer the top aqueous layer to a clean tube and discard the bottom organic layer.
6. Add 100 µL chloroform, vortex for 15 s and spin for 5 min as in **step 4**.
7. Transfer the top aqueous layer to a clean tube and discard the bottom organic layer.
8. Add 0.1 vol 3 *M* sodium acetate, pH 6.0, and 2.5 vol 100% ethanol. Precipitate the RNA at –70°C for 30 min.
9. Spin for 10 min at maximum speed in a microcentrifuge to recover the RNA. Discard the supernatant taking care not to dislodge the RNA pellet.
10. Wash the pellet by adding 1 mL 70% (v/v) ethanol and spinning for 5 min. Take off the supernatant and allow the pellet to dry under vacuum. Resuspend the RNA pellet in an appropriate amount of RNase-free dH_2O.

3.2. First Strand cDNA Synthesis

1. Program the thermocycler as follows: 65°C for 5 min, 37°C for 90 min, 95°C for 5 min.
2. Add the following components to a RNase-free microcentrifuge tube (*see* **Note 3**):

Component	Amount required	Final concentration
M-MLV reaction buffer (10X)	2 µL	1X
dNTPs (25 m*M*)	2 µL	2.5 m*M*
Anchor primer (200 µ*M*) (*see* **Notes 4–6**)	1 µL	10 µ*M*
RNA (0.1–1.0 µg)		
dH_2O	to 18 µL	

3. Place the tubes in the thermocycler and start the program. After 10 min of incubation at 37°C, interrupt the program and add the following components:

Component	Amount required	Final concentration
RNase inhibitor (20–40 U/μL)	1 μL	20–40 U
M-MLV RT (50 U/μL)	1 μL	50 U

4. Mix well and allow the thermocyler program to continue.
5. When the program is complete, add 60 μL sterile dH_2O to each cDNA synthesis reaction.
6. The cDNA can either be used for PCR immediately or stored at –20°C.

3.3. PCR Amplification

1. Program the thermocycler as follows: 40 cycles of 94°C for 30 s, 40°C for 2 min, 72°C for 30 s, and followed by 1 cycle at 72°C for 5 min.
2. Add the following components to a sterile microfuge tube (*see* **Note 7**):

Component	Amount required	Final concentration
cDNA	2 μL	
PCR reaction buffer (10X)	2 μL	1X
$MgCl_2$ (50 m*M*)	0.4 μL	1 m*M*
dNTPs (25 μM)	1.6 μL	2 μ*M*
Anchor primer (200 μ*M*)	1 μL	10 μ*M*
Abitrary primer (50 μ*M*) (*see* **Note 8**)	1 μL	2.5 μM
^{35}S-dATP (>1000 Ci/mmol)	0.5 μL	
Taq DNA polymerase (5 U/ μL)	0.2 μL	1 U
dH_2O	to 20 μL (*see* **Note 9**)	

3. Vortex and centrifuge briefly to bring the contents to the bottom of the tube.
4. Overlay the samples with 100 μL (1 drop) mineral oil (*see* **Note 10**).
5. Place the tubes in the thermocycler and start the program.

3.4. Denaturing Polyacrylamide Gel Electrophoresis

1. Using a tissue, spread 2 mL Gel Slick™ on the glass plate that contains the buffer reservoir (the back plate). Allow the Gel Slick to dry and then wipe using a clean tissue. Do not put Gel Slick on the glass plate that forms the front plate of the apparatus (*see* **Note 11**).

2. Assemble the sequencing gel apparatus according to the manufacturer's instructions (*see* **Note 12**).

3.4.1. Preparation of a 5% Polyacrylamide Gel

1. Prepare the gel mix by adding the following components to a clean 50-mL beaker:

Component	Amount required	Final concentration
Urea	21 g	7 *M*
Long Ranger™ gel solution (50X)	5 mL	5X
10X TBE	6 mL	1.2X
dH_2O	to 50 mL	

3. Allow the urea to dissolve completely (heat gently if necessary). Once dissolved, keep the beaker containing the gel mix on ice.
4. To a sterile universal bottle on ice, add 4 mL gel mix, 7 µL TEMED and 40 µL 25% APS. Quickly mix and pour 2 mL of the seal mix down each side of the glass plates using a P1000 Gilson pipet. Allow the seal to set across the bottom of the plates for 5–10 min (*see* **Note 13**).
5. Once the seal has set, add 25 µL TEMED and 250 µL 10% APS to the remaining 46 mL gel mix. Mix thoroughly.
6. Fill a 20-mL syringe with the gel mix and, holding the gel apparatus at a 45° angle, pour the gel mix down one side of the plates, therefore avoiding the formation of air bubbles.
7. When the space between the glass plates is completely full with gel mix, lay the plates down flat on the bench, then raise the top end 2 in. using a pipet tip box as support.
8. Insert the sharkstooth comb 5 mm so that the flat edge is facing the bottom of the gel. Apply any remaining gel mix to the exposed part of the comb. Allow the gel to set in this position for at least 1 h.

3.4.2. Electrophoresis and Autoradiography

1. Remove the masking tape from the bottom of the gel plates and the excess gel mix from the top of the gel plates using a spatula.
2. Mount the gel cassette onto the sequencing apparatus according to the manufacturer's instructions.
3. Take out the sharkstooth comb and wash off any excess gel mix with water.
4. Prepare sufficient 1X TBE running buffer (*see* **Note 14**) to fill both anode and cathode chambers by diluting 10X TBE stock.

5. Flush the top of the gel with 1X TBE running buffer using a P1000 Gilson pipet.
6. Reinsert the comb so that the flat edge is uppermost and the "teeth" are 2–3 mm into the gel, forming the wells into which the samples are loaded (*see* **Note 15**).
7. Prerun the gel at 50 W until the temperature of the glass plates is 50°C (usually 30 min–1 h).
6. Add 4 µL gel loading buffer to each sample and mix (*see* **Note 16**).
7. Incubate the samples at 80°C for 2 min to denature the DNA fragments, then cool on ice to avoid renaturation.
8. Flush the wells with 1X TBE running buffer using a pasteur pipet or P1000 Gilson pipet.
9. Load 6 µL of each sample onto the gel and run at 45–50 W constant power until the xylene cyanol dye is 10 cm from the bottom of the gel (usually ~2 h) (*see* **Note 17**). The power should be adjusted if necessary so that the temperature of the glass plates is maintained at ~50°C.
10. Following electrophoresis, turn off the power supply and allow the plates to cool briefly. Discard the running buffer and remove the clamps from each side of the gel cassette. Lay the gel cassette flat on the bench so that the glass plate containing the buffer reservoir is uppermost. Separate the glass plates so that the gel remains on the plate that forms the front part of the apparatus.
11. Cut a piece of Whatman 3MM filter paper so that it is just larger than the glass plate.
12. Place the filter paper over the gel and press down to encourage the gel to stick to the filter paper and remove any air bubbles. Peel the filter paper and gel from the plate and lay flat on the bench so that the gel is uppermost.
13. Cover the gel with Saran Wrap and wipe with a tissue to remove any air bubbles.
14. Dry the gel at 80°C under vacuum for 1 h.
15. Remove the Saran Wrap and expose the gel to X-ray film (*see* **Note 18**) in a light-tight cassette. Make a number of cuts around the edges of the dried gel and film to allow accurate orientation following autoradiography (*see* **Note 19**).

3.5. Elution and Reamplification of DNA Fragments

3.5.1. Elution of DNA Fragments

1. Once bands of interest have been identified (*see* **Fig. 2**) align the autoradiogram with the dried gel. Tape the film and gel together once

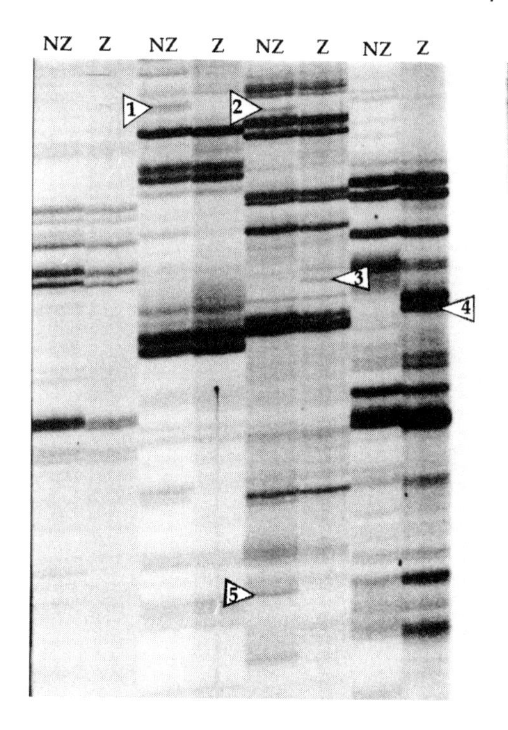

Fig. 2. A typical differential display autoradiogram is shown. The arrows are pointing to five differential bands (labeled 1–5) that were eluted from the dried down gel and reamplified (*see* **Fig. 3**). Z = abscission zone, NZ = nonzone.

the correct alignment is achieved. Using a pencil, mark the areas on the reverse of the gel that contain the bands of interest.

2. Using a scalpel, cut out the gel slices and rehydrate them for 20 min in a microfuge tube containing 150 µL dH_2O.
3. Boil the gel slices/dH_2O for 20 min, then spin for 10 min at maximum speed in a microcentrifuge.
4. Transfer the supernatant to a clean tube and make the vol up to 200 µL with dH_2O.
5. Add 0.1 vol of 3 *M* sodium acetate (pH 5.2), 10 µL glycogen (10 mg/mL) (*see* **Note 20**) and 2.5 vol 100% ethanol.
6. Allow the DNA to precipitate at –20°C for at least 1 h.
7. Spin for 10 min at maximum speed in a microcentrifuge to recover the DNA. Discard the supernatant, taking care not to dislodge the DNA/glycogen pellet.

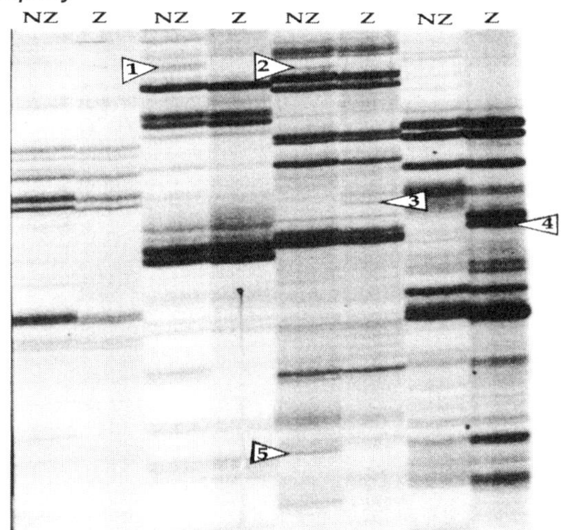

Fig. 3. A 4% (w/v) agarose gel on which the five reamplified bands (labeled 1–5) had been loaded.

8. Wash the pellet by adding 1 mL ice cold 85% (v/v) ethanol and spin for 5 min. Take off the supernatant and allow the pellet to dry under vacuum. Resuspend the pellet in 10 µL dH_2O and store at –20°C.

3.5.2. Reamplification of DNA Fragments

1. Add the following components to a sterile microfuge tube:

Component	Amount required	Final concentration
Eluted DNA	4 µL	
PCR reaction buffer (10X)	4 µL	1X
$MgCl_2$ (50 m*M*)	0.8 µL	1 m*M*
dNTPs (250 µ*M*)	3.2 µL	20 µ*M*
Anchor primer (200 µ*M*)	2 µL	10 µ*M*
Abitrary primer (50 µ*M*)	2 µL	2.5 µ*M*
Taq DNA polymerase (5 U/µL)	0.4 µL	2 U
dH_2O	to 40µL	

3. Vortex and centrifuge briefly to bring the contents to the bottom of the tube.
4. Overlay the samples with 100 µL (1 drop) mineral oil (*see* **Note 10**).
5. Place the tubes in the thermocycler and start the program. Use the same program as for the differential display PCR (**Subheading 3.3.**).
6. Run 20 µL each sample on a 4% low melting point agarose gel (*see* **Fig. 3**) (*see* **Note 21**). If there are no visible products on the gel, a

second round PCR amplification can be performed using the same PCR conditions and 4 µL of the first round PCR reaction as the template.

4. Notes

1. If stored for long periods of time this solution (10X TBE) can form a precipitate. To overcome this problem, either prepare a less concentrated stock solution, e.g., 5X TBE or prepare <1 L of 10X solution.
2. It is important that the RNA is of a high quality (Absorbance ratio $A_{260/A280}$ = 1.8) and free of contaminating genomic DNA. When RNA is badly contaminated with genomic DNA, the display pattern becomes largely independent of reverse transcription. The RNA is therefore treated with DNase to remove any genomic DNA.
3. It is advisable to include a minus reverse transcriptase (-RT) control that contains all the components required for first strand cDNA synthesis except M-MLV reverse transcriptase. Following the subsequent differential display PCR reaction and polyacrylamide gel electrophoresis, any bands that appear in the samples and in the -RT control can be eliminated. These PCR products are the result of amplification of contaminating genomic DNA, not cDNA produced by reverse transcription of the mRNA.
4. As an alternative to the oligo-dT anchor primers, specific or degenerate oligonucleotide primers to your gene of interest may be used to prepare first-strand cDNA.
5. In our experience, the degenerate anchor primers (1–4) give higher background than the nondegenerate primers (5–12). However, this can be used to your advantage as a way of decreasing the number of false positives; for example, a differential band produced using anchor primer 3, which contains a penultimate degenerate 3' base (i.e., anchor primer 3 contains a mixture of 3 different primers). dT(12)GG, dT(12)AG and dT(12)CG) should also be produced when using anchor primer 7 (dT(12)GG) or anchor primer 11 (dT(12)CG). Any differential bands that do not appear in both samples should be avoided. Similarly, this rule applies to primer sets 1 and 8 or 12; 2 and 5 or 9; 4 and 6 or 10.
6. Numerous articles in the literature have reported different anchor primer design. One such article by Liang et al. *(3)* proposed the use of one-base anchor primers, which allow subdivision of the mRNA into three populations, therefore reducing the number of reverse transcription reactions required for each RNA sample.

7. In order to reduce the number of false positives, it is advisable to perform the PCR reactions in duplicate or triplicate so that bands that do not appear in both duplicated samples can be eliminated.
8. As an alternative to the arbitrary primer in the PCR amplification reaction, specific or degenerate oligonucleotide primers to your gene of interest may be used in conjunction with the primer (oligo-dT anchor, degenerate or specific to your gene of interest) that was used in the first strand cDNA synthesis reaction.
9. The PCR amplification step also works well when scaled down in proportion to 10 µL.
10. Mineral oil can be omitted if your thermocycler has a heated lid.
11. Gel Slick prevents the gel from sticking to the glass plate. It is applied to one glass plate only so that when the plates are separated after electrophoresis, the gel will stick to the front plate only and can then be transferred easily to Whatman 3MM filter paper.
12. In our experience, it is easier to seal the gap at the bottom of the assembled glass plates with masking tape. This is removed before electrophoresis.
13. This mixture constitutes a quick setting polyacrylamide slurry. It is poured into the gap between the glass plates (2 mL for each side) and allowed to settle at the bottom. The mix sets quickly and forms a plug, therefore reducing the risk of any leakages when the main gel is poured.
14. If required, in order to enhance the appearance of the higher molecular weight (upper) bands, a TBE concentration of 1.2X in the gel and 0.6X in the running buffer is recommended.
15. It is sometimes difficult to replace the comb into the gel. If this is so, apply a small amount of grease to the "teeth." This will also help to prevent leakages between the wells when the gel is loaded.
16. The mineral oil can be removed if necessary by pipeting off as much as possible, or by a single extraction with 100 µL chloroform.
17. Xylene cyanol migrates at a slower rate (approx 250 bp in a 5% polyacrylamide gel) than bromophenol blue (approx 65 bp in a 5% polyacrylamide gel) and is turquoise in color.
18. BioMax-MR (Kodak) film has maximum resolution and sensitivity to ^{35}S- labeled gels.
19. The exposure time is determined by the number of counts per second as measured by holding a Geiger counter next to the dried gel. As a general rule, 20–50 cps requires 24 h exposure, 10–20 cps requires 48 h exposure, and <10 cps requires >48 h exposure.

20. The concentration of DNA eluted from the gel is usually very low so glycogen is added to act as a coprecipitant. Glycogen forms a large white pellet following centrifugation, most of which is dissolved during the 85% (v/v) ethanol wash. Any residual glycogen will not interfere with the reamplification reaction.
21. NuSieve GTG low melting point agarose finely resolves nucleic acid fragments of <1 kb and can distinguish fragments as small as 8 bp. Since a high proportion of bands eluted from differential display gels are <500 bp, a 4% low melting point agarose gel allows more accurate size determination of these low mol wt PCR fragments than normal agarose gels (*see* **Fig. 3**).

References

1. Liang, P. and Pardee, A. B. (1992) Differential display of eukaryotic messenger RNA by means of the polymerase chain reaction. *Science* **257,** 967–971.
2. Liang, P., Bauer, D., Averboukh, L., Warthoe, P., Rohrwild, M., Muller, H., Strauss, M., and Pardee, A. B. (1995) Analysis of altered gene expression by differential display. *Methods Enzymol.* **254,** 304–321.
3. Liang, P., Zhu, W., Zhang, X., Guo, Z., O'Connell, R.P., Averboukh, L., Wang, F. and Pardee, A. B. (1994) *Nucleic Acids Res.* **22,** 5763–5764.

3

Abscisic Acid

ABA Immunoassay and Gas Chromatography/Mass Spectrometry Verification

M. K. Walker-Simmons, Patricia A. Rose, Lawrence R. Hogge, and Suzanne R. Abrams

1. Introduction

1.1. ABA Immunoassays

Development of monoclonal antibodies (MAbs) that specifically recognize the plant hormone, (+)-(*S*)-abscisic acid (ABA) have enabled the development of ABA immunoassays *(1,2)*. ABA measurement with these immunoassays is efficient, sensitive, and suitable for large numbers of samples (reviewed in **ref. *3***).

This chapter describes methods for ABA immunoassay by an indirect enzyme-linked immunosorbent assay (ELISA) using MAbs. Before immunoassay, sample extraction and clean-up is required to remove substances that may interfere with the immunoassay. The indirect ELISA assay consists of first incubating samples (plant tissue extracts or ABA standards) with MAb to ABA. The MAb binds ABA in the samples, leaving varying levels of free MAb. Then, the amount of free MAb is measured in the remaining steps of the ELISA assay. ABA content in the original samples is inversely proportional to the amount of free MAb measured in the ELISA assay.

From: *Methods in Molecular Biology, vol. 141: Plant Hormone Protocols*
Edited by: G. A. Tucker and J. A. Roberts © Humana Press Inc., Totowa, NJ

Thus, ABA content can be calculated based on standard curves of known ABA standards.

When the indirect ELISA is used to measure ABA in a plant tissue for the first time, the assay results should be verified by an independent method. A method for verifying the ELISA assay results is described using gas chromatography/mass spectrometry (GC/MS) and a deuterated internal standard.

Antibodies used in the ELISA assay are specific for (+)-(*S*)-ABA, the natural form of ABA in plants. Different grades of ABA are commercially available, including racemic cis,trans-ABA, which contains an equal amount of (+)-(*S*)- and (-)-(*R*)-ABA. ABA MAbs used for ELISA assays should not recognize (-)-(*R*)-ABA or cross-react significantly with ABA metabolites or other components in the sample ***(4)***.

ELISA methods using MAbs have been used to assay ABA in plant tissue during growth and environmental stress ***(5–8)***; or mutant screening ***(9)***, and for genotype comparison ***(10)***. Other effective ABA immunoassay methods have been developed, including radioimmunoassay (RIA) ***(1,2,11)***, fluoroimmunoassay (FIA) ***(12,13)*** and an enzyme-amplified immunoassay ***(14,15)***. An ABA immunoassay detection kit is also available (Sigma, St. Louis, MO, product no. PGR-1 or Agdia, Elkart, IN).

Advantages of the indirect ABA ELISA immunoassay include:

1. It requires small amounts of plant tissue (can be used to measure ABA levels in 5–10 mg samples, depending on endogenous ABA content).
2. The assay is sensitive and detects small amounts (pg) of ABA.
3. A large number of samples can be assayed efficiently and economically.
4. No radioactive materials are required for routine assay.
5. No expensive instrumentation is required.

2. Materials

2.1. ABA Extraction

1. Plant tissue: Freshly harvested tissue should be frozen rapidly using liquid nitrogen and then stored at –20°C.

2. Grinding of plant tissue: Micropestles and microtubes or prechilled mortar and pestle.
3. Extracting methanol: 100% methanol (analytical grade) containing 0.5 g/L citric acid monohydrate (Baker) and 100 mg/L butylated hydroxytoluene (BHT) (Calbiochem). (Mention of a specific product named by the United States Department of Agriculture does not constitute an endorsement and does not imply a recommendation over other suitable products.)
4. Disposable 14 mL (17 × 100 mm) polypropylene test tubes.

2.2. Sample Clean-Up

1. 100, 70, and 62.5% extracting methanol. Prepare 70 and 62.5% extracting methanol by adding distilled water. For example, to prepare 62.5% extracting methanol; add 62.5 mL extracting methanol to 37.5 mL water.
2. Sep-Pak C_{18} cartridges (Millipore). Our laboratory does not reuse the C_{18} cartridges.
3. Disposable syringes (3 mL).
4. Disposable 14-mL (17 × 100 mm) polypropylene test tubes.
5. [^{3}H]ABA (Amersham, Arlington Heights, IL).

2.3. ABA Immunoassay (Indirect ELISA)

1. Immulon 4 microtitration plates, 96 flat-bottom wells of protein-binding polystyrene (Dynatech Laboratories, Inc., Chantilly, VA, catalog no. 011-010-3855). Plates can lose protein binding capability with long-term storage (*see* **Note 1**).
2. Monoclonal antibody to ABA (Agdia Inc., Elkhart, IN, Cat. no. PDM 09347/0096). Prepare a stock antibody solution by adding 2 mg ABA MAb to 135 mL TBS containing 0.2% (0.27 g) bovine serum albumin (BSA, Sigma). Prepare 3 mL aliquots (sufficient for one assay plate) and store frozen at –20°C (*see* **Note 2**).
3. Repeating, variable volume pipeter (such as an Eppendorf Repeater Pipettor) for repetitive dispension of solutions containing proteins. Use this pipeter for antibody and ABA-BSA conjugate solutions. Do not reuse reservoir tips.
4. Tris buffered saline (TBS): 6.05 g Tris-HCl, 0.2 g $MgCl_2(6H_2O)$, 8.8 g NaCl/L, pH 7.8.
5. TBS-Tween: TBS plus 0.5 mL/L Tween 20, pH 7.8.

6. TBS washing buffer: TBS-Tween plus 1 g/L BSA, pH 7.8.
7. (+)-(*S*)-ABA standard: Prepare a stock solution of (+)-(*S*)-ABA (Sigma) in dimethyl sulfoxide (DMSO) by adding 1.0 mg (+)-ABA/0.1 mL DMSO. Mix well. Cap tightly and store in the dark at 4°C. Prepare a diluted stock solution of ABA in TBS for ABA calibration standards of 1×10^6 pg/0.1 mL TBS. Dilute (+)-ABA to 1×10^4 pg/0.1 mL TBS, and then prepare ABA standards for each ELISA assay plate consisting of 167, 125, 100, 83, 50, 25, 10, or 5 pg/0.1 mL TBS. A volume of 0.4 mL of each ABA standard is required per assay plate. Diluted ABA standards should be prepared fresh at least every month. Store all ABA solutions at 4°C (dark). *See* **Note 3**.
8. ABA-$C_{4'}$-BSA conjugate (modified from Quarrie and Galfre; ***16***). *See* **Note 4**.
9. 0.05 *M* $NaHCO_3$, pH 9.6 (2.1 g/500 mL distilled water).
10. Rabbit antimouse IgG alkaline phosphatase conjugate (Sigma, cat. no. A-1902).
11. *P*-Nitrophenyl phosphate (5 mg Sigma "104" phosphatase substrate tablets, cat. no. 104-105).
12. ELISA plate reader.

2.4. ABA Immunoassay Verification (Gas Chromatography/Mass Spectrometry)

1. Glass containers, such as Reacti-Vials (0.3 mL, Pierce Chemical Co., Rockford, IL), are convenient for the mass spectrometry standards and samples, because they seal well. Glass, rather than plastic vials, should be used for all ABA standards and samples for GC/MS, to avoid contamination from plasticizers.
2. Use isopropanol/glacial acetic acid (95/5) as the extraction solvent.
3. Deuterated ABA standard for addition to plant samples as an internal standard: Prepare a stock solution of 3',5',5',7',7',7'-hexadeuteroabscisic acid (ABA-d_6), ***(17)*** (one source is the Plant Biotechnology Institute, National Research Council of Canada, Saskatoon, Canada) of 2.0 ng/µL by dissolving 0.50 mg ABA-d_6 in 2.5 mL methanol and then diluting with distilled water to 250 mL in a volumetric flask. ABA standards used in mass spectrometry can be racemic.
4. ABA/ABA-d_6 calibration standards for mass spectrometric analysis of the ABA methyl esters: First, prepare a stock solution of ABA-d_6

at 1 µg/µL by adding 1.00 mg ABA-d_6 to 1.00 mL methanol, and then prepare a dilute stock solution of 1 ng/µL methanol. Similarly, prepare a stock solution of ABA at 1 µg/µL using methanol, and then dilute stock solutions in methanol to 1 and 0.1 ng/µL. Use the stock solutions to prepare a series of ABA/ABA-d_6 standards so that each vial contains 100 ng ABA-d_6 and 1, 5, 10, 50, or 100 ng of ABA. Remove the solvent by evaporation using a gentle stream of nitrogen gas at room temperature. For mass spectral analysis, the ABA and ABA-d_6 standards are converted to the methyl esters as follows: Treat the solvent-free standards with excess (0.3 mL) of fresh ethereal diazomethane (to be undertaken by a technically qualified person; *see* **Note 6**). Let samples stand for 15 min. Remove the residual diethyl ether with nitrogen gas. Add 100 µL ethyl acetate to each standard, and ensure the vial is well sealed.

3. Methods

3.1. ABA Extraction

This protocol has been used to extract ABA from seeds and shoot tissue of wheat and barley. Alternate extraction methods include extraction into boiling water ***(18)*** and crude extraction of multiple samples for mutant screening ***(9)***.

Controls should be run to monitor ABA yield and recovery through the extraction and clean-up procedures. These controls should include adding radiolabeled-ABA ([^{3}H]ABA, Amersham, Arlington Heights, IL) or a known amount of (+)-ABA to crude tissue samples. Quantify recovery by measuring radioactivity in a liquid scintillation counter and by assaying for (+)-ABA using the indirect ELISA assay (described in **Subheading 3.3.**).

1. Weigh (fresh weight) and then rapidly freeze plant tissue in liquid nitrogen. Store samples at –20°C. The amount of sample required depends on the ABA content of the tissue. A 20–50-mg sample of wheat seed embryos is convenient for assay, though even smaller amounts of tissue can be assayed.
2. Freeze-dry (lyophilize) samples.
3. Grind samples to a powder. Embryos can be ground to a powder using a small amount of liquid nitrogen and a small mortar and pestle. Weigh sample (dry weight).

4. Transfer powdered sample to a test tube (17 x 100 mm) containing a small stirring bar. Add extracting methanol at a ratio of 10 mg dry tissue/1.0 mL extracting methanol. Stir suspension in sealed tube for at least 20 h at 4°C in the dark.
5. Centrifuge samples at 1500*g* for 10 min.
6. Collect supernatant. This supernatant sample from wheat embryos can be assayed directly using the indirect ELISA assay described in **Subheading 3.3.** Control experiments and verification analysis *(8)* have shown that further clean-up of wheat embryo extracts is not necessary. However, many plant tissue samples require further clean-up using Sep-Pak cartridges or equivalent procedures (i.e., other solid-phase extraction cartridges, filters, or high performance liquid chromatography [HPLC]) to remove substances that interfere with the ELISA assay. Sep-Pak cartridges adapted for vacuum manifold devices are convenient for multiple samples. Sample clean-up procedures are described in **Subheading 3.2.** Whenever the indirect ELISA method is used to measure ABA in a new type of plant tissue, the immunoassay results should be first verified by GC/MS before routine use of the indirect ELISA. The procedure for GC/MS verification is described in **Subheading 3.4.**

3.2. Crude Sample Clean-Up

1. Adjust crude extract supernatant samples to 70% methanol by diluting samples at a ratio of 200 µL extract sample to 800 µL of 62.5% extracting methanol. Mix well.
2. Prewash the Sep-Pak C_{18} cartridge by attaching a cartridge filter to a syringe barrel and place in a upright position. Aliquot 2 mL of 100% extracting methanol into the syringe and push methanol slowly through the Sep-Pak cartridge with the syringe plunger. Remove the cartridge from the syringe and then remove the syringe plunger. Then reattach the Sep-Pak cartridge to the syringe. Load 1 mL of 70% extracting methanol into the syringe and push slowly through the cartridge filter with the syringe plunger. Remove the cartridge from the syringe and then remove the plunger. Reattach the Sep-Pak cartridge to the syringe barrel.
3. Load the plant extract sample diluted to 70% extracting methanol (1–2 mL total volume) into the syringe attached to the equilibrated Sep-Pak cartridge. Very slowly push the samples through into a clean 14-mL test tube.

4. Remove the Sep-Pak cartridge from the syringe and then remove the plunger. Reattach the syringe to the cartridge. Wash the cartridge with 1 mL 70% methanol. Slowly push the sample through the column with the plunger and collect in the 14-mL test tube. Cap eluted sample and store at 4°C in the dark. A control should be conducted to check that no ABA is left on the Sep-Pak cartridge by washing the cartridge with 2–4 mL more of 70% methanol.
5. Dry the eluted samples in a Speed-Vac Concentrator (Savant, Farmingdale, NY) to dryness. Store dried samples at –4°C. Occasionally, a sample cannot be completely dried by Speed-Vac. Dry as completely as possible and then continue with immunoassay.

3.3. ABA Indirect ELISA Assay

This protocol describes (+)ABA measurement by indirect ELISA using one 96-well assay plate. Accurate ABA measurement of a tissue sample requires that ABA standards be measured on each plate. Plant samples should be diluted sufficiently so that ABA levels are calculated from A_{405} values that are within the linear range of the ABA standard curve. Tissue samples should be assayed at two or three dilutions with three replicates of each dilution. This means that ABA measurement of one tissue sample requires 6–9 wells of the 96-well assay plate. The upper (row A) and lower (row H) rows of the 96-well plate are often not used because these wells tend to produce more variable results.

3.3.1. Day 1

All solutions should be kept on ice or at 4°C.

1. Sample preparation. Resuspend dried samples (original dry weight, approx 20 mg) in 100 µL methanol and then add 900 µL TBS. Mix well. Prepare serial dilutions, such as 1/5 and 1/10 in TBS (*see* **Note 5**). A volume of 400 µL is required for each diluted sample.
2. Prepare 400 µL of each ABA standard (167, 125, 100, 83, 50, 25, 10, or 5 pg/100 µL TBS). Keep ABA standards at 4°C (dark).
3. Prepare MAb solution. Add 3.0 mL MAb stock to 9.0 mL TBS. Mix gently. Using an Eppendorf repeater pipet, add 400 µL MAb to 400 µL of each plant sample or ABA standard. Mix by vortexing gently.

Incubate samples at room temperature for 30 min (dark). Mix again, and incubate overnight (dark) at 4°C.

4. Add 10 μL of ABA-BSA conjugate to 16 mL of 0.05 *M* $NaHCO_3$, pH 9.6. Add 200 μL conjugate solution to each well of the 96-well assay plate using a repetitive pipeter. (exclude rows A and H). Cover and incubate at 4°C (dark).

3.3.2. Day 2

1. Dump the ABA-BSA conjugate from the 96-well plate. Slap the inverted plate on paper towels to remove as much remaining solution as possible. Wash the plate with TBS washing buffer by filling each well completely using a multiple pipet or a wash bottle and then dumping the wash solution. Repeat the washing three times, leaving the wash solution in the wells for 10 min during the final wash cycle. Hit the inverted plate on paper towels to remove remaining solution. Do not allow plates to dry out during the assay. Dump the final wash solution only when solutions are prepared for the next assay step.
2. Add 200 μL of samples or ABA standards incubated with MAb to three replicate wells. Cover and incubate for 2 h (dark) at room temperature.
3. Prepare rabbit antimouse IgG alkaline phosphatase solution by adding 25 μL conjugate to 16 mL TBS. Dump samples from the 96-well plate and wash three times with TBS washing buffer. Dump the final wash solution. Add 200 μL rabbit antimouse alkaline phosphate conjugate solution to each well using a repetitive pipeter. Incubate for 2 h (dark) at room temperature.
4. Prepare *p*-nitrophenyl phosphate substrate solution by dissolving four tablets (5 mg each) in 20 mL 0.05 *M* $NaHCO_3$. Dump samples from the 96-well plate and wash three times with TBS washing buffer. Dump the final wash solution. Add 200 μL *p*-nitrophenyl phosphate substrate solution to each well. Incubate approx 2 h (dark, room temperature).
5. Measure absorbance at 405 nm of samples in the 96-well plate using an ELISA plate reader. Incubate until A_{405} of control samples containing no ABA is approx 1.0. Record A_{405} of samples with an ELISA plate reader. The amount of ABA in the plant extract samples is inversely proportional to the A_{405} of the samples.
6. Average the replicate ABA sample results. Conduct a linear regression analysis using the ABA standard results. Calculate the amount

of ABA in the plant extract samples using the coefficient of the ABA standard curve for the plate. Plant sample extracts should always be diluted sufficiently so that A_{405} values fall within the linear portion of the ABA standard curve. Calculate ABA content in sample extracts using only sample results that are within the linear range of the ABA standard curve.

3.4. Verification of ELISA Using GC/MS

This protocol has been used to analyze ABA from wheat seeds, to verify the ELISA method ***(19)***. In the GC/MS method, the methyl esters of ABA are analyzed. The method below describes the preparation of samples suitable for analysis by a mass spectrometry laboratory. An internal standard is added at the beginning of the extraction process. The internal standard is an isotopically labeled analog of ABA with six deuterium atoms replacing protons and has a mass 6 Daltons greater than ABA. In the GC/MS analysis, racemic deuterated ABA can be used, because both the natural (+)- and unnatural (-)-isomer give the same spectrum. Solvent used for extraction of ABA from plant tissue should not contain preservatives, such as butylated hydroxytoluene, which interfere with the mass spectral analysis. Use a quantity of tissue that contains between 10 and 100 ng ABA. For best mass spectrometry results, the quantity of internal standard added should be approximately the same as the amount of endogenous ABA expected to be present in the plant sample. The final volume of the sample for mass spectral analysis should be such that the concentration of ABA-d_6 is 1.0 ng/μL.

1. Rapidly freeze plant tissue in liquid nitrogen, and then freeze dry sample, grind to a powder, and sieve.
2. Transfer 10 mg of powdered sample to an Eppendorf tube and add 1.0 mL isopropanol/glacial acetic acid (95/5). Add internal standard 20 ng ABA-d_6 (10 μL of stock solution). Shake using a Thermomixer at room temperature for 18 h in the dark.
3. Centrifuge at 1500*g* for 10 min.
4. Collect supernatant, transfer to a clean Eppendorf tube, and dry the sample in a Speed-Vac Concentrator to dryness.
5. Transfer to a Reacti-Vial with methanol and remove the solvent with a gentle stream of nitrogen gas.

6. Add 0.3 mL freshly prepared diazomethane (to be undertaken by a technically qualified person; *see* **Note 6**). Let stand for 15 min, and evaporate to dryness under a gentle stream of nitrogen gas.
7. Just before mass spectral analysis, take up in 20 µL ethyl acetate to generate a 1 ng/µL ABA-d_6 solution, and seal well to avoid evaporation. Mass spectrometry equipment and experiment parameters are given in **Note 7**.
8. Once the spectra have been obtained, the quantity of ABA in the plant samples to which a known amount of ABA-d_6 has been added can be calculated. Calculate the ratio of the response for the m/z 190 ion (from Me ABA) to the response for the m/z 194 ion (from Me ABA-d_6). From the ratio, calculate the amount of ABA in the sample per 1.0 ng of ABA-d_6 from the calibration curve. Use sample results only that fall with the linear range of the ABA/ABA-d_6 calibration curve.

4. Notes

1. Microtiter plates can lose protein-binding capability with long-term storage of 6 mo or more. Do not prewash microtiter plates or reuse plates. Old (washed) plates can be used as covers.
2. Monoclonal antibody stock solution can be stored frozen for months, but can lose crossreactivity with long-term storage. Batches of MAb can vary in ABA specificity. When a new MAb stock solution is prepared, the ELISA assay should be run with ABA standards using both the old and new MAb stock solutions. If the linear range of the ABA standard curve changes significantly, the ratio of MAb stock solution to TBS-Tween can be adjusted.
3. If (+)-(*S*)-ABA is not available, standards of racemic (±)-ABA consisting of (+)- and (-)-ABA can be used to measure ABA levels. The MAb will only recognize the natural hormone, (+)-ABA, which is 50% of racemic ABA.
4. Preparation of ABA-$C_{4'}$-BSA conjugate (modified from ***16***).
   a. First prepare ABA-4'-*p*-aminobenzoyl hydrazone. (Wear safety glasses for all conjugate preparation steps.) Recrystalize *p*-aminobenzoyl hydrazide (ABH) by adding 50 mg *p*-aminobenzoyl hydrazide (4-aminobenzoic hydrazide, Aldrich, cat. no. A4,190-9) to 35 mL methanol. Carefully heat methanol to approx 50–55°C until ABH just dissolves. Do not overheat. Add more methanol if

necessary to completely dissolve ABH. Recrystallize by placing a beaker containing ABH solution on ice. Centrifuge (1000*g*, 5 min) and discard supernatant. Dry ABH crystals under vacuum or by directing a stream of N_2 over the crystals.

b. Add 15.5 mg recrystallized ABH to 3 mL methanol containing 1 *M* glacial acetic acid. Prepare methanol containing 1 *M* glacial acetic acid by adding 0.54 mL glacial acetic acid to 9.43 mL methanol. Mix and then purge with N_2. Add 13.5 mg (+)-(*S*)-ABA. Racemic (±)-ABA can be substituted (it is less expensive). Purge solution with N_2. Leave the test tube at room temperature (in dark) for 3 d. Purge 1–2 times/d with N_2 and keep the test tube capped. The solution will develop a yellow color.
c. Dry sample under N_2 (in partial darkness). This may take several hours. A yellow oil residue is left.
d. Partition residue between borate (Baker) buffer, pH 9.0, and ethyl acetate (Baker). Add 3 mL 0.1 *M* borate buffer, pH 9.0, and 3 mL ethyl acetate to the test tube containing the residue. Mix and allow the two layers to form. Discard ethyl acetate layer. Repeat by adding 3 mL ethyl acetate to the remaining 0.1 *M* borate buffer, mix, and discard the ethyl acetate layer again. Use glass test tubes for all steps involving ethyl acetate.
e. Adjust aqueous (borate) phase to pH 4.5 h. Partition with ethyl acetate by adding 3 mL ethyl acetate to the aqueous phase. Mix. Remove and save the ethyl acetate phase. Repeat two times, each time pooling the ethyl acetate phases. Dry combined ethyl acetate extracts under N_2 in the dark, leaving a yellow ABA-4'-*p*-aminobenzoyl hydrazone residue. Store under N_2 in the dark at 4°C.
f. Prepare the following solutions and cool to 0°C on ice:
   i. Add 160 mg BSA (Sigma) to 8 mL 0.1 *M* borate buffer, pH 9.2, and then dialyze overnight in 0.25 *M* borate buffer, pH 9.2.
   ii. Dissolve ABA-4'-*p*-aminobenzoyl hydrazone residue in 200 µL methanol, and then add 800 µL 0.1 *N* HCl.
   iii. $NaNO_2$ (Sigma) (75 mg/3 mL distilled water).

   Add 70 µL $NaNO_2$ dropwise and very slowly to 500 µL ABA-4'-*p*-aminobenzoyl hydrazone. Incubate for 20 min as diazotization takes place. (Store remaining ABA-4'-*p*-aminobenzoyl hydrazone at 4°C in the dark.) Add the solution dropwise slowly to 1.5 mL BSA in borate buffer. Incubate for 16 h at 4°C (dark). Dialyze against borate buffer, pH 9.0, for 3 d (dark). Replace the dialysis solution 1–2 times/d with fresh borate buffer.

g. Collect the dialyzed ABA-BSA conjugate. Repeat the process by adding the remaining 500 µL ABA-4'-*p*-aminobenzoyl hydrazone to the dialyzed ABA-BSA conjugate solution. Then, add 70 µL $NaNO_2$ slowly, dropwise, and incubate for 20 min. Add diazotized solution to 1.5 mL BSA dialyzed against borate buffer. Incubate overnight at 4°C in the dark. Then, dialyze against PBS buffer (1.5 L) with several changes of fresh buffer for 2 d. PBS buffer: 1.44 g Na_2HPO_4, 0.24 g KH_2PO_4, 8.0 g NaCl, 0.2 g KCl/L, pH 7.5.

h. Aliquot 100 µL ABA-BSA conjugate into clean test tubes and freeze. Store frozen until used. (The conjugate is destroyed by repeated thawing and freezing.) Batches of ABA-$C_{4'}$-BSA conjugate can vary. When a new batch of ABA-BSA conjugate is prepared, several aliquot volumes (i.e., 5, 10, or 20 µL of ABA-BSA conjugate) should be tested with ABA standards in the ELISA assay along with an aliquot of the older ABA-BSA conjugate.

5. Interfering substances in plant samples or from the extraction media can interfere with the accuracy of the ELISA assay ***(20)***. A concentration of 5% (v/v) or less of methanol in the diluted plant samples used for ELISA assay is recommended. When the assay is used for a new type of plant tissue, controls should be conducted measuring the yield and accuracy of the assay. Internal controls with known amounts of ABA should always be conducted to check on the yield and accuracy of ABA measurement.
6. Diazomethane is highly toxic and explosive. Safety precautions must be taken, including wearing safety glasses, using a fumehood for all preparations and manipulations, and carefully following instructions provided with the reagent Diazald, which is used to make diazomethane (Technical Information Bulletin AL-180, Aldrich, Milwaukee, WI).
7. Parameters for GC/MS analyses: In our laboratory GC/MS for ABA content is carried out using a Micromass Trio 2000 GC/MS system in the electron impact ionization mode. Sample injections of 1.0 µL (in the splitless injection mode) are made onto a 60 m × 0.32 mm DB-23 (J&W Scientific, Folsom, CA) fused silica column. The injection port temperature used is 270°C. The initial column temperature is set to 70°C, raised ballistically to 180°C, and then programmed at 4°C/min to 260°C.

Mass spectral analysis is done in the selected ion monitoring (SIM) mode. The fragment ions monitored for ABA methyl ester (Me ABA) are m/z 134, 162, and 190 whereas those monitored for the internal standard (Me ABA-d_6) are m/z 138, 166, and 194. The intensities of the ions m/z 190 (Me ABA) and 194 (Me ABA-d_6) are compared for quantitation whereas the remaining ions monitored are used to identify the analyte. Each ion is monitored for 80 ms with a settling time of 20 ms.

A calibration curve is generated from the mass spectra of the standards, plotting on the *x*-axis the known amount of ABA per nanogram of ABA-d_6 in each standard and on the y-axis plotting the ratio of the response for the m/z 190 ion (from Me ABA) to the response for the m/z 194 ion (from Me ABA-d_6). A typical calibration curve has an r^2 of 0.988.

Acknowledgment

We thank Joan Sesing-Lenz, Lynn D. Holappa, and Guy Birkenmeier for helping develop ELISA methods and Garth Abrams and Doug Olson for contributions to the extraction and mass spectrometry developments.

References

1. Mertens, R., Deus-Neumann, B., and Weiler, E. W. (1983) Monoclonal antibodies for the detection and quantitation of the endogenous plant growth regulator, abscisic acid. *FEBS Lett.* **160,** 269–272.
2. Quarrie, S. A., Whitford, P. N., Appleford, N. E. J., Wang, T. L., Cook, S. K., Henson, I. E., and Loveys, B. R. (1988) A monoclonal antibody to (*S*)-abscisic acid: its characterisation and use in a radioimmunoassay for measuring abscisic acid in crude extracts of cereal and lupin leaves. *Planta* **173,** 330–339.
3. Walker-Simmons, M. and Abrams, S. R. (1991) Use of ABA immunoassays, in *Abscisic Acid: Physiology and Biochemistry*, (Davies, W. J. and Jones, H. G., eds.), Bios Scientific, Oxford, UK, pp. 53–61.
4. Walker-Simmons, M. K., Reaney, M. J. T., Quarrie, S. A., Perata, P., Vernieri, P., and Abrams, S. R. (1991) Monoclonal antibody recognition of abscisic acid analogs. *Plant Phyisol.* **95,** 46-51.
5. Holappa, L. D. and Blum, U. (1991) Effects of exogenously applied ferulic acid, a potential allelopathic compound, on leaf growth, water utilization, and endogenous abscisic acid levels of tomato, cucumber, and bean. *J. Chem. Ecol.* **17,** 865–886.

6. Norman, S. M., Poling, S. M., and Maier, V. P. (1988) An indirect enzyme-linked immunosorbent assay for (+)abscisic acid in *Citrus*, *Ricinus*, and *Xanthium* leaves. *J. Agric. Food Chem.* **36,** 225–231.
7. Ross, G. S., Elder, P. A., McWha, J., Pearce, D., and Pharis, R. P. (1987) The development of an indirect enzyme linked immunoassay for abscisic acid. *Plant Physiol.* **85,** 46–50.
8. Walker-Simmons, M. (1987) ABA levels and sensitivity in developing wheat embryos. *Plant Physiol.* **84,** 61–66.
9. Walker-Simmons, M., Kudrna, D. A., and Warner, R. L. (1989) Reduced accumulation of ABA during water stress in a molybdenum cofactor mutant of barley. *Plant Physiol.* **90,** 728–733.
10. Tahara, M., Guenzi, A. C., Read, J. J., Carver, B. R., and Johnson, R. C. (1991) Quantification of abscisic acid in wheat leaf tissue by direct enzyme immunoassay. *Crop Sci.* **31,** 1185–1189.
11. Vernieri, P, Perata, P., Armellini, D., Bugnoli, M., Presentini, R., Lorenzi, R., Ceccarelli, N., Alpi, A., and Tognoni, F. (1989) Solid phase radioimmunoassay for the quantitation of abscisic acid in plant crude extracts using a new monoclonal antibody. *J. Plant Physiol.* **134,** 441–446.
12. Banowetz, G. M. (1989) Hybrid hybridoma cell lines which express anti-zeatin riboside, anti-abscisic acid, and hybrid antibodies. *Plant Physiol.* **91,** 144–147.
13. Banowetz, G. M., Hess, J. R., and Carman, J. G. (1994) A monoclonal antibody against the plant growth regulator, abscisic acid. *Hybridoma* **13,** 537–541.
14. Harris, M. J. and Outlaw, W. H. , Jr.(1990) Histochemical technique: a low-volume, enzyme-amplified immunoassay with sub-fmol sensitivity. Application to measurement of abscisic acid in stomatal guard cells. *Physiol. Plant.* **78,** 495–500.
15 Harris, M. J., Outlaw, W.H., Jr., Mertens, R., and Weiler, E. W. (1988). Water-stress-induced changes in the abscisic acid content of guard cells and other cells of *Vicia faba* L. leaves as determined by enzyme-amplified immunoassay. *Proc. Natl. Acad. Sci. USA* **85,** 2584–2588.
16. Quarrie, S. A. and Galfre, G. (1985) Use of different hapten-protein conjugates immobilized on nitrocellulose to screen monoclonal antibodies to abscisic acid. *Anal. Biochem.* **151,** 389–399.
17. Bonnafous, J.-C., Fonzes, I., and Mousseron-Canet, M. (1971) Syntheses d'acide abscisique radioactif. II Marquage de tritium. *Bull. Soc. Chim. France* 4552–4554.

18. Loveys, B. R. and van Dijk, H. M. (1988) Improved extraction of abscisic acid from plant tissue. *Aust. J. Plant Physiol.* **15,** 421–427.
19. Tavakkoi-Afshari, R., Hucl, P., Rose, P. A., and Abrams, S. R. (1998) Temporal variation in abscisic acid content of four wheat genotypes. 8th International Symposium on Pre-Harvest Sprouting in Cereals, Association of Cereal Research, Federal Centre for Cereal, Potato and Lipid Research, Detmold, Germany, *Abstract* **18,** June 2–5.
20. Belefont, H. and Fong, F. (1989) Abscisic acid ELISA: organic acid interference. *Plant Physiol.* **91,** 1467–1470.

4

Auxin Analysis

Els Prinsen, Stijn Van Laer, Sevgi Öden, and Henri Van Onckelen

1. Introduction

Plant hormones are present and active in minor concentrations in plant tissue. To allow quantification in a restricted amount of material, i.e., protoplasts, chloroplasts, seedlings, seeds, buds, or apical root- and stem regions, a sensitive analytical technique is a prerequisite. During the last two decades, procedures for phytohormone analysis as well as the available hardware have improved substantially. Fluorimetry is widely used for the analysis of indole-3-acetic acid (IAA) because it is more specific and therefore more sensitive than ultraviolet detection (for a review *see* **ref. *1***). Good detection limits for derivatized IAA are obtained in Electron Impact (EI^+) *(2,3)* or negative ion chemical ionization (NICI) GC-MS* *(4,5)*.

IAA can be analyzed by GC-MS after methylation *(5)*, heptafluorobutyryl acylation *(6)*, *N*-trifluoroacetyl or *N*-heptafluorobutyryl derivatives (*see* **ref. *4***), trichloroethyl esterification (*see* **ref. *4***), pentafluorobenzyl esterification *(4,5,7–9)*, or trimethylsilylation *(2)*. During these derivatization procedures, special attention should be paid to avoid degradation of the derivatized IAA in contact with trace amounts of water. Special storage under desic-

**See* end of chapter for definitions of acronyms.

From: *Methods in Molecular Biology, vol. 141: Plant Hormone Protocols*
Edited by: G. A. Tucker and J. A. Roberts © Humana Press Inc., Totowa, NJ

cated conditions are in most cases advised. The stability of these derivatives was analyzed and compared by Schneider et al. ***(5)***. The method described by Epstein and Cohen ***(4)*** provides quantitative yields of pentafluorobenzyl (PFB)-esters without anhydrous conditions and uses reagents that are stable under normal storage conditions. Moreover, because of the presence of the pentafluorobenzyl group, the molecule yields additional electron capture characteristics. This allows the analysis of the negative ion [M^-] after electron capture in a chemical ionization mode. This ionization mode is far more specific, and therefore results in a lower background noise. Many undetected impurities will not interfere with the analysis. Netting and Millborrow ***(8)*** adapted the derivatization procedure to produce PFB-IAA to allow simultaneous derivatization and analysis of IAA and ABA. We adapted the derivatization procedure as described by Epstein and Cohen ***(4)*** for the following reasons:

1. It has an easy derivatization protocol.
2. This derivative is stable for several months.
3. Anhydrous conditions are not required.
4. ABA and GAs are derivatized in a similar manner, which allows the simultaneous analysis of the latter phytohormones during one GC-MS run.

The first reports on the analysis of IAA-conjugates and catabolites ***(10,11)*** used LC-frit fast atomic bombardment (FAB)-MS, but usually IAA-conjugates are analyzed after alkaline hydrolysis ***(12)***. We described the use of LC(ESI)-MS/MS for the analysis of indole compounds ***(13)***. The electrospray ionization process by itself is extremely soft, optimizing the probability of obtaining molecular mass information ***(14,15)***, and is therefore very suitable for the analysis of phytohormones, such as auxins and cytokinins. In combination with conventional LC, a detection limit of 0.1 pmol IAA injected on-column was achieved under single reaction monitoring conditions. Despite the structural analogy between the different compounds analyzed simultaneously, the unique diagnostic transitions for each individual compound used for single reaction monitoring (SRM) channel enabled fast analysis under less chromatographic resolution ***(13,16)***.

Mass sensitivity can be improved 230 times by miniaturization in LC instrumentalization as well as in sample size as compared to conventional HPLC *(17)*. In combination with GC analysis, microscale analysis of IAA from a few milligrams to <1 mg starting material should be obtained by optimizing the extraction volumes *(2,3)*. Although Ribnicky et al. *(3)* showed the analysis of IAA in <1 mg starting material by EI GC-MS after methylation, an ideal sensitivity is obtained in combination with a more selective detection, such as NICI, or double-focusing high resolution mass spectrometry *(2)*.

The method described in this chapter is an update of the procedure published earlier in **ref.** ***18***.

2. Materials

2.1. Preparation of the Diethylaminoethyl (DEAE)-Sephadex A25 (Anion Exchange Column, Formate Conditions)

1. Dissolve 100 g dry DEAE Sephadex A_{25} in 500 mL distilled water for swelling following the manufacturer's guidelines. Never use a magnetic stirrer because this might damage column particles.
2. Rinse the swollen DEAE with 3 L 1 *M* NH_4-formate. A separation funnel is very useful for this purpose.
3. Rinse the column material with distilled water until pH 6.0 is reached.
4. Check if the pH of the column material is between 5.0 and 7.0 before use. If not, then rinse with additional water or 50% methanol.
5. Always use freshly prepared column material.
6. Pour 2 mL of column material in a 5-mL syringe containing a polypropylene frit.

2.2. C_{18} Column Cartridges

The reversed phase (RP)- C_{18} cartridges used are single-use prepacked 1 or 0.5 mL cartridges (Bond-Elut by Varian, Harbor City, CA ; Macherey-Nagel GmbH & Co, Düren, Germany; Baker Inc., Phillipsburg, NJ or IST Ltd., Mid Glamorgan, UK). Prerinse the cartridges with 10 mL technical grade ethanol for conditioning the column matrix followed by 20 mL distilled water and 10 mL of the solvent to be used. Although these columns are single use, they can easily be reconditioned by rinsing with 10 mL acetone. We could recycle these cartridges over 50 times without reducing the capacity.

2.3. Extraction and Purification of Free IAA

1. Liquid nitrogen.
2. –70 or –20°C freezer.
3. 80% Methanol (HPLC grade).
4. Phenyl-$^{13}C_6$-IAA (Cambridge Isotope Lab., Andover, MA) *see also* **Note 1**.
5. Mortar and pestle.
6. MeOH-resistant centrifuge tube or, for small sample volumes, an Eppendorf microtube.
7. High-speed centrifuge (Centrikon T-124, Kontron, Milan, Italy) or scintered glass filter. For small amounts an Eppendorf centrifuge is appropriate.
8. C_{18} cartridge.
9. Distilled water.
10. 0.0–14.0 pH indicator strips.
11. 50% MeOH (HPLC grade).
12. DEAE-Sephadex A_{25} cartridge.
13. 6% formic acid.
14. $NaSO_4$ cartridge (isolute sodium sulfate drying cartridge, IST Ltd 2.5 g, or MN Chromafix 1200-Dry, 1.2 g) (optional).
15. Diethylether.
16. 1 or 3 mL reaction vial.
17. Evaporator using a nitrogen stream, i.e., a Pierce Reacti-Vap Evaporator (Pierce, Rockford, IL).

2.4. Purification of IAA-Amino Acid Conjugates

1. 10 mL Snap cap reaction vial.
2. 7 *N* NaOH.
3. Water-saturated nitrogen circuit (obtained by applying a water trap in a nitrogen flow).

4. Incubator at 100°C.
5. Ice.
6. Distilled water.
7. 2 *N* HCl.
8. 0.0–14.0 pH indicator strips.
9. 0.05 *N* HCl.
10. 80% MeOH (HPLC grade).

2.5. Purification of IAA-Sugar Conjugates

1. 10 mL Snap cap reaction vial.
2. 1 *N* NaOH.
3. Water-saturated nitrogen circuit.
4. Distilled water.
5. 2 *N* HCl.
6. 0.0–14.0 pH indicator strips.
7. 0.05 *N* HCl.
8. 80% MeOH (HPLC grade).

2.6. Synthesis of the Pentafluorobenzyl Derivative of IAA (PFB-IAA)

1. A 1- or 3-mL gas-tight reaction vial.
2. Evaporator using a nitrogen stream, i.e., a Pierce Reacti-Vap Evaporator.
3. Acetone p.a.
4. 1-Ethylpiperidine.
5. α-Bromopentafluorotoluene (= pentafluorobenzyl bromide).
6. Incubator at 60°C, i.e., a Pierce Reacti-Therm Heating module.
7. Ethylacetate p.a.
8. H_2O (milliQ or HPLC grade).
9. Methanol 100% (HPLC grade).

2.7. Synthesis of the IAA Methyl Ester (Me-IAA)

The methylation of IAA is appropriate when analysis will be performed by LC-MS/MS. Electron impact (EI) GC-MS of the IAA methyl ester is possible and on a routine basis applied in several laboratories. However, the larger the diagnostic ion, the higher the specificity will be. Therefore, we prefer the PFB-IAA derivative for GC-MS analysis.

1. A 1- or 3-mL gas-tight reaction vial.
2. Acidified methanol (2 drops of fuming HCl added to 100 mL 100% methanol).
3. Ethereal diazomethane (freshly synthesized following **refs.** ***18*** and ***19***.
4. Nitrogen stream for drying samples, i.e., a Pierce Reacti-Vap Evaporator. Vacuum evaporation (i.e., using a Speed-Vac) is not appropriate as the IAA methyl ester is volatile.
5. 100% Methanol (HPLC grade).
6. 20% Methanol (HPLC grade methanol, MilliQ water).
7. MilliQ water.
8. Eppendorf centrifuge.

2.8. GC-MS Analysis

1. Chemical ionization energy moderator: methane or ammonium gas (Air Liquide).
2. Column: J&W (Folsom, CA) DB-XLB 15 m, id 0.25 mm, film thickness 0.25 μm (J&W).
3. Labbase or Masslynx (VG Micromass Ltd., Manchester, UK) GC-MS software.
4. GC-coupled mass spectrometer with chemical ionization or electron impact source (VG Micromass Trio 2000).
5. Hewlett Packard 5890 Series II gas chromatograph (Hewlett Packard, Avondale, PA) and HP 6890 injector with a Gerstel Sys3 cooled large volume injection system (Gerstel GmbH, Mülheim, Germany).

2.9. LC (+)ESI SRM MS/MS Analysis of IAA

1. Kontron 465 (Kontron Instruments, Milan, Italy) injector equipped with a 10-, 25-, or 100-μL sample loupe, depending on the sample volume injected.
2. Injector-driven 10-port multifunctional valve with microelectric two-position valve actuator (Valco Instrument Co. Inc., Houston, TX).
3. Through-hole 0.5-mm SS precolumn filter unit (JOUR Research, Onsala, Sweden)
4. Prodigy 5 μm OD83 100 Å 30 × 1 mm (Phenomenex, Torrane, CA) precolumn.

5. HPLC pump (Kontron 325).
6. 0.01 *M* NH_4OAc pH 7.
7. MeOH (HPLC grade).
8. Hypersil 5 µm C_8 BDS 100 x 1 mm id (Alltech) analytical column, Prodigy 5 µm OD83 100 Å 50 × 1 mm, or a 30 × 1 mm (Phenomenex). The Alltech columns have as an extra advantage that column filter frits can easily be replaced.
9. Two HPLC pumps (Kontron 422) equipped with 0.01–2 mL/min pump heads in master-slave configuration. This is a cheaper alternative for an HPLC gradient pump with a maximal 0.01–2 mL/min pump head volume.
10. Biocompatible PEEK Mixing Tee (JOUR Research) as a low-volume solvent mixing device to couple two Kontron 422 in gradient mode.
11. Quattro II mass spectrometer (Micromass Ltd. Manchester, UK) equipped with an electrospray interface and a megaflow probe (Micromass Ltd.).

3. Methods

3.1. Extraction

1. After harvest, immediately freeze the excised plant material in liquid nitrogen. Keep samples at –70°C until analysis (*see also* **Note 2**).
2. Homogenize the plant material in liquid nitrogen using a mortar and pestle. For calli or soft, nonlignified tissue *see* **Note 3**. For the extraction of plant or bacterial cells *see* **Note 4**. For the extraction of protoplasts, *see* **Note 5**. For small amounts of sample *see* **Note 6**.
3. Extraction can easily be obtained immediately in a centrifuge tube.
4. Add 80% methanol to 5 mg–2 g plant tissue (9 mL/gram fresh weight).
5. Add the $^{13}C_6$-IAA internal tracer. The amount used (10–1000 pmol) depends on the amount of starting material. The exact amount added should be registrated.
6. IAA is extracted overnight from the tissue at –20°C in darkness.

3.2. Purification of Free IAA

1. Remove cell debris by centrifugation : 24,000*g*, 4°C, 15 min. For alternatives *see* **Note 7** and **8**.

2. Pass the supernatant at 80% MeOH over a C_{18} cartridge to remove chlorophylls.
3. Rinse the column after application of the sample with an additional 10 mL of 80% MeOH.
4. Collect the effluent, containing IAA and pool with the cartridge's 10 mL wash volume.
5. Dilute the sample with water until 50% MeOH is obtained.
6. Purify the sample using a DEAE-sephadex A_{25} cartridge (prerinsed with 50% MeOH; control the column's pH [5.0–6.0]).
7. Rinse the DEAE cartridge after application of the sample with an additional 75 mL of 50% MeOH.
8. Prerinse a C_{18} cartridge sequentially with 5 mL ethanol, 10 mL water, and finally 5 mL 6% formic acid.
9. Couple the C_{18} cartridge underneath the DEAE-sephadex cartridge.
10. Elute IAA from the DEAE-sephadex with 20 mL 6% formic acid. IAA is retained on the C_{18} matrix. The effluent of the C_{18} can be discarded.
11. Pass one syringe of air through the C_{18} cartridge.
12. Couple a $NaSO_4$ cartridge underneath the C_{18} for on-line drying of the ethereal sample (optional).
13. Elute IAA from the C_{18} cartridge with 1 mL diethylether. Collect the diethylether immediately in a reaction vial.
14. Discard the water phase, if any, underneath the ether phase.
15. Evaporate the ether phase under a nitrogen stream. Cap the vial upon storage.

3.3. Purification of IAA-Conjugates

Although IAA-conjugates can be analyzed in their native form by LC-MS ***(10,11)***, we prefer to analyze IAA-conjugates after alkaline hydrolysis ***(12)***. This is merely because heavy labeled internal tracers for the different IAA-conjugates are not easily available. When alkaline hydrolysis is performed, **Note 9** should be taken into account. The amount of IAA analyzed after alkaline hydrolysis corresponds to the total amount of IAA (free + conjugated). Therefore, the amount of free IAA as analyzed following **Subheading 3.2.** should be subtracted.

3.3.1. IAA-Amino Acid Conjugates

1. Take one-half to one-tenth of the original extract after stage **Subheading 3.1.** and transfer to a 10-mL snap cap reaction vial.
2. Add NaOH until a final concentration of 7 *N* NaOH is obtained. *See also* **Note 10**.
3. Apply a water-saturated nitrogen circuit by applying a water trap in a nitrogen flow.
4. Flush the vial to obtain a water-saturated nitrogen atmosphere.
5. Incubate the vial at 100°C for 3 h.
6. Cool down the sample on ice.
7. Dilute one-half with distilled water and titrate the sample with 2 *N* HCl until pH 2.5 is obtained.
8. Desalt the acidified sample on a C_{18} cartridge that is equilibrated with 0.05 *N* HCl.
9. Rinse the cartridge with an additional 10 mL 0.05 *N* HCl.
10. Elute the cartridge with 80% MeOH.
11. Dilute with distilled water to obtain a final MeOH concentration of 50%.
12. Continue at **step 6** in **Subheading 3.1.**

3.3.2. IAA-Sugar Conjugates

1. Take one-half to one-tenth of the original extract after **Subheading 3.1.** and transfer to a reaction vial.
2. Add NaOH until a final concentration of 1 *N* NaOH is obtained. *See also* **Note 10**.
3. Apply a water-saturated nitrogen circuit by applying a water trap in a nitrogen flow.
4. Flush the vial to obtain a water-saturated nitrogen atmosphere.
5. Incubate the vial at ambient temperature for 1 h.
6. Dilute one-half with distilled water and titrate the sample with 2 *N* HCl until pH 2.5 is obtained.
7. Desalt the acidified sample on a C_{18} cartridge that is equilibrated with 0.05 *N* HCl.
8. Rinse the cartridge with an additional 10 mL 0.05 *N* HCl.
9. Elute the cartridge with 80% MeOH.
10. Dilute with distilled water to obtain a final MeOH concentration of 50%.
11. Continue at **step 6** in **Subheading 3.1.**

3.4. Derivatization of IAA

3.4.1. Pentafluorobenzyl Ester of IAA (PFB-IAA)

This derivatization procedure is essential for the NICI GC-MS analysis of IAA (prior to **Subheading 3.5.1.**). Different derivatization methods are described for the analysis of IAA, as summarized in **Subheading 1.** We have chosen the most stable and easiest derivatization, which in addition allows chemical ionization analysis (because of the electron capture characteristics of the pentafluorobenzyl group) *(4)*.

Warning: Most of the derivatization mixtures are toxic; work in a fume-hood during the entire derivatization procedure.

1. Transfer the sample in a reaction vial and dry entirely.
2. Dissolve in 50 µL acetone p.a.
3. Add 1 µL 1-ethylpiperidine and 5 µL α-bromopentafluorotoluene (= pentafluorobenzyl bromide) and mix vigorously.
4. Incubate at 60°C for 45 min.
5. Add 300 µL ethylacetate and mix vigorously.
6. Add 300 µL H_2O (milliQ or HPLC grade) and mix vigorously.
7. Remove and discard the lower water phase.
8. Repeat from **step 6** on.
9. Dry the ethylacetate fraction under a nitrogen stream.
10. Dissolve the sample in 20 µL methanol 100% (HPLC grade) (*see also* **Note 12**).

3.4.2. IAA Methyl Ester (Me-IAA)

This derivatization procedure is performed prior to LC (+)ESI SRM MS/MS analysis of IAA. Although methylation is not essential, this manipulation improves the detection limit by a factor of 100 (prior to **Subheading 3.5.2.**). Methylation is performed following Schlenk and Gellerman *(19)*.

Warning: The diazomethane is toxic. Work in a fume-hood during these steps. Take care that the vial containing the sample is resistant to ether, i.e., glass.

1. Transfer the sample in a glass reaction vial and dry entirely. Avoid water condensation inside the reaction vial at any circumstance throughout all further steps. (*see also* **Note 11**).

2. Dissolve in 100 μL acidified methanol.
3. Add ethereal diazomethane until a slight yellow color remains for 10 min at room temperature.
4. Dry the sample under a nitrogen stream. Avoid vacuum evaporation (i.e., Speed-Vac) because the IAA methyl ester is volatile.
5. Dissolve in 50 μL 100% methanol (HPLC grade).
6. Transfer the sample to an LC-MS insert.
7. Dry the methanol fraction under a nitrogen stream.
8. Dissolve in 10 μL 20/80, v/v, methanol/H_2O.
9. Add 30 μL MilliQ water until a final MeOH concentration of 5% is reached.
10. Mix vigorously.
11. Spin down the sample in the insert. (*see also* **Note 12**).

3.5. Analysis

Two different analysis methods are outlined here. GC is on a routine base used for the analysis of IAA. ESI LC-MS/MS have, however, several advantages in terms of selectivity, especially when several indole compounds are analyzed in parallel (*see also* **Note 13**.)

3.5.1. NICI GC-MS Analysis of IAA

3.5.1.1. GC-MS Settings

1. Injection volume: 1–10 μL.
2. GC oven gradient: 2 min 175 °C followed by a linear gradient at a temperature rate of 15°C min to 300°C and finally 2 min at 300°C.
3. Solvent delay (2 min).
4. Injector temperature: temperature gradient from 150 to 325°C in 2 min.
5. Source temperature: 140°C by constant filament emission.
6. Source vacuum pressure: 7–8 10^{-5} mbar.
7. GC carrier gas: helium (Air Liquide), He flow: 1.5 mL/min.
8. Ionization potential: 35 eV.
9. Ionization current: 100 μA.
10. Negative chemical ionization mode.
11. Analysis by selective ion monitoring (SIM).
12. Diagnostic ion IAA-PFB: m/z at 174.
13. Diagnostic ion $^{13}C_6$-IAA-PFB: m/z at 180.

14. Dwell time: 0.01 s.
15. Inter channel delay: 0.01 s.

3.5.1.2. Integration and Calculation

All data were processed by Labbase or Masslynx (VG Micromass) software.

The endogenous concentration of IAA is calculated from the spectra obtained using the following equation:

$$(\text{area}_x/\text{area}_y) = (x/y) \qquad (1)$$

where : area_x = peak integration value for the specific diagnostic ion for the nonlabeled IAA, area_y = peak integration value for the specific diagnostic ion for the $^{13}C_6$-labeled IAA, x = endogenous IAA content initially present in the tissue analyzed, and y = amount of $^{13}C_6$-labeled IAA added to the extract before purification.

Correct for the 180/174 ratio of the $^{13}C_6$-IAA tracer. The detection limit of this method is 0.5 fmol injected.

3.5.2. LC/LC (+)ESI SRM MS/MS Analysis of IAA

On-line sample concentration can be obtained by on-column focusing or a column switching setup, the latter having an additional on-line purification.

3.5.2.1. Chromatographic Conditions

1. Precolumn: Prodigy 5 µm OD83 30 × 1 mm (Phenomenex).
2. Analytical column: Hypersil 5 µm C8 BDS 150 × 1 mm id column (Alltech, Deerfield, IL), or a Prodigy 5 µm OD83 100 or 50 × 1 mm (Phenomenex).
3. Injection volume: 25 µL.
4. Solvent A: 0.01 *M* NH_4OAc pH 6.6.
5. Solvent B: Methanol 100%.
6. Solvent delivery for loading precolumn: 100% solvent A.
7. Solvent delivery for backflushing: 4 min sample loading at 5/95, B/A; from 1 to 3 min: linear gradient from 5/95 to 90/10 B/A.
8. Flow rate: 100 µL/min.

9. Pump systems: two Kontron 422 pumps equipped with 0.01–2 mL/min pump heads in master-slave configuration using a Biocompatible PEEK Mixing Tee (JOUR Research) as a low-volume solvent mixing device.
10. The effluent was directly introduced into the MS source at a flow rate of 100 μL/min.

3.5.2.2. Mass Spectrometry Conditions

A crossflow counter electrode (Micromass) was used to avoid excessive contamination of the source for routine analysis of biological samples. The mass spectrometer was tuned using a 10^{-5} *M* Me-IAA solution dissolved in 100% methanol.

1. Source temperature: 80°C.
2. Nebulizing gas flow: 20 L/h.
3. Drying gas flow: 400 L/h.
4. Capillary voltage: +3.5 kV.
5. Cone voltage: 28 V.
6. Collision gas: Argon, P_{AR} of 4 10^{-3} mbar
7. Collision energy 20 eV.
8. Quantification: single reactant monitoring (SRM) of the $[MH]^+$ ion and the appropriated product ion (190→130 for IAA-Me, 196→136 for the $^{13}C_6$-IAA-Me).
9. Dwell time: 0.05 s.
10. Inter channel delay: 0.01 s.
11. Span: 0 amu.
12. The effluent of the analytical column is directly introduced into the MS source at a flow rate of 100 μL/min.
13. Sample loading by a Kontron 325 pump for 6.5 min.
14. After 6.5 min sample loading, the precolumn is backflushed.
15. While sample 1 is introduced to the analytical column, a second sample was loaded on a parallel precolumn.

3.5.2.3. Integration and Calculation

All mass spectra is background subtracted and smoothed once. All data is processed by Masslynx software. Endogenous IAA concentrations are calculated using **Eq. 1**. The detection limit of this method is 1 fmol injected.

4. Notes

1. Different heavy-labeled tracers for IAA are available. A tracer should always be chosen in terms of a high enrichment and a large amount of labeled sites (at least two or more). Deuterated tracers are shown to be less stable and exchange in alkaline or acidic conditions may occur.
2. –70°C is preferred for the storage of samples before analysis. However, for storage <3 mo, –20°C is also acceptable.
3. For calli or soft, nonlignified tissue, the extraction is performed by diffusion without prior homogenization.
4. Plant or bacterial cells are lysed by sonication using a Vibra Cell VCX400 High Intensity Untrasonic Processor (Sonics and Materials Inc., Danbury, CT) with a stepped microtip to allow small volume (200 μL–10 mL) processing. Sonicate under a pulsed regimen of 5 s (5 s sonication–5 s rest) for a 4 min total time span. Use a power of 400 W at 40% gain. Continuously cool the samples on ice during processing.
5. Extract protoplasts in 500 μL 80% methanol in an Eppendorf tube.
6. Small amounts of sample are easily extracted in an Eppendorf microtube after homogenization using an Eppendorf size conical pestle.
7. Cell debris can also be removed by filtration using a scintered glass filter under vacuum.
8. Small amounts of plant material that are extracted in an Eppendorf microtube are centrifuged at full speed using an Eppendorf centrifuge.
9. If IAA-conjugates are hydrolyzed under alkaline hydrolysis, data obtained for the total IAA fraction are contaminated because of degradation of IAN into IAA ***(20)***. The use of an $^{13}N_1$ -IAN tracer (gift from Dr. N. Ilic, University of Maryland) is essential to calculate the sample-specific correction factor measuring the IAA+1 diagnostic ion.
10. Concentration of the sample prior to alkalization is advised in case the original extract volume is too large.
11. Avoid traces of water during methylation.
12. For additional purification after derivatization (before **Subheading 3.5.**), a C_{18} cartridge or, if large amounts of samples are processed, C_{18} 96-well extraction plates are useful. The small eluting volume of the latter makes immediate elution (100 μL 100% MeOH) in LC or GC inserts possible. Sample processing can either be achieved manu-

ally using a 96-well vacuum manifold or it can be automated using a number of commercially available sample processors. The apolar characteristics of the IAA-derivative allows loading of a sample in neutral watery conditions and subsequent elution in 100% MeOH.

13. In case the chemical ionization mode (CI, NICI) is not available, IAA can easily be analyzed by electron impact (EI) GC-MS after pentafluorobenzylation or methylation: diagnostic ion IAA-PFB or IAA-Me: m/z at 130; diagnostic ion $^{13}C_6$-IAA-PFB or $^{13}C_6$-IAA-Me: m/z at 136 ***(11)***. Source temperature: 270°C by constant filament emission. Source vacuum pressure: 7–8 10^{-5} mbar. Ionization potential: 70 eV. Ionization current: 250 µA.We prefer to analyze the PFB-IAA using a chemical ionization interface although electron impact is also perfectly acceptable. Because of the specificity of CI (correlated to the rare ion-capture characteristics), the background noise is much more reduced, resulting in improved sensitivity.
14. Sometimes we experienced contaminating compounds, although with a retention time slightly different from PFB-IAA, which could result in altered peak shape for IAA. We experienced these interfering compounds as less stable. In this case, storage of derivatized samples at –20°C for at least 1 wk before GC-MS analysis always resulted in a improvement of the chromatography for PFB-IAA without significant recovery losses. The PFB-IAA derivative is stable for several months without any significant degradation.

Acknowledgments

E. P. and H. V. O. are respectively Postdoctoral Fellow and Research Director of the Fund for Scientific Research Flanders (F. W. O.). S. V. L. is a Research Assistant of the Flemish Institute for Scientific Technological Research in Industry (I. W. T.). Dr. N. Ilic (University of Maryland) synthesized the $^{13}N_1$ -IAN tracer.

Abbreviations:

CI, Chemical ionization; EI, electron impact; ESI, electrospray ionization; GC-MS, gas chromatography coupled mass spectrometry; IAA, indole-3-acetic acid; LC (ESI)-MS/MS, liquid chromatography coupled to electrospray tandem mass spectrometry; LC, liquid chromatography; LC/LC, liquid chromatography in combination with column switch; LC-MS, liquid chromatography coupled mass spectrometry; $[M^-]$, molecular ion after electron capture; Me, methyl ester; $[MH]^+$, protonated molecular ion; NICI, negative ion chemical ionization; PFB, pentafluorobenzyl ester; SIM, selected ion monitoring; SRM, single reactant monitoring.

References

1. Kobzar, E. F. and Serdyuk, O. P. (1993) High performance liquid chromatography for analysis of auxins. Biochemistry (New York-English translation of *Biokhimiya*), **58,** 33–40.
2. Edlund, A., Eklöf, S., Sundberg, B., Moritz, T., and Sandberg, G. (1995) A microscale technique for gas chromatography-mass spectrometry measurements of picogram amounts of indole-3-acetic acid in plant tissues. *Plant Physiol.* **108,** 1043–1047.
3. Ribnicky, D. M., Cooke, T. J., and Cohen, J. D. (1998) A microtechnique for the analysis of free and conjugated indole-3-acetic acid in milligram amounts of plant tissue using a benchtop chromatograph-mass spectrometer. *Planta* **204,** 1–7.
4. Epstein, E. and Cohen, J. D. (1981) Microscale preparation of pentafluorobenzyl esters: electron-capture gas chromatographic detection of indole-3-acetic acid from plants. *J. Chromatogr.* **209,** 413–420.
5. Schneider, E. A., Kazakoff, C. W., and Wightman, F. (1985) Gas chromatography-mass spectrometry evidence for several endogenous auxins in pea seedling organs. *Planta* **165,** 232–241.
6. Pilet, P.-E. and Saugi, M. (1985) Effect of applied and endogenous indol-3-yl-acetic acid on maize root growth. *Planta* **164,** 254–258.
7. Markham, G., Lichty, D. G., and Wightman, F. (1980) Comparative study of derivatization procedures for the quantitative determination of the auxin, phenylacetic acid, by gas chromatography. *J. Chromatogr.* **192,** 429–433.
8. Netting, A. G. and Millborrow, B. V. (1988) Methane chemical ionization mass spectrometry of the pentafluorobenzyl derivatives of abscisic acid, its metabolites and other plant growth regulators. *Biomed. Environ. Mass Spectrom.* **17,** 281–286.
9. Netting, A. G. and Duffield, A. M. (1985) Positive and negative ion methane chemical ionization mass spectrometry of amino acid pentafluorobenzyl derivatives. *Biomed. Mass Spectrom.* **12,** 668–672.

10. Östin, A., Moritz, T., and Sandberg, G. (1992) Liquid chromatography/mass spectrometry of conjugates and oxidative metabolites of indole-3-acetic acid. *Biol. Mass Spectrom.* **21,** 292–298.
11. Östin, A., Catalá, C., Chamarro, J., and Sandberg, G. (1995) Identification of glucopyranosyl-β-1,4–glucopyranosyl-β-1-*N*-oxindole-3-acetyl-*N*-aspartic acid, a new IAA catabolite, by liquid chromatography/tandem mass spectrometry. *J. Mass Spectrom.* **30,** 1007–1017.
12. Bialek, K. and Cohen, J. D. (1989) Quantitation of Indole-3-acetic acid conjugates in bean seeds by direct tissue hydrolysis. *Plant Physiol.* **90,** 398–400.
13. Prinsen, E., Van Dongen, W., Esmans, E., and Van Onckelen, H. (1997) HPLC Linked electrospray tandem mass spectrometry: a rapid and reliable method to analyze indole-3-acetic acid metabolism in bacteria. *J. Mass Spectrom.* **32,** 12–22.
14. Burlingame, A. L., Boyd, R. K., and Gaskell, S. J. (1996) Mass spectrometry. *Anal. Chem.* **68,** 599R–651R.
15. Niessen, W. M. A. and Tinke, A. P. (1995) Review: liquid chromatography-mass spectrometry general principles and instrumentation. *J. Chromatogr. A* **703,** 37–57.
16. Prinsen, E., Redig, P., Van Dongen, W., Esmans, E. L., and Van Onckelen, H. (1995) Quantitative analysis of cytokinins by electrospray tandem mass spectrometry. *Rapid Comm. Mass Spectrom.* **9,** 948–953.
17. Prinsen, E., Van Dongen, W., Esmans, E. L., and Van Onckelen, H. (1998) Micro and capillary LC-MS/MS : a new dimension in phytohormone research. *J. Chromatogr. A* **826,** 25–37.
18. Prinsen, E., Redig, P., Strnad, M., Galis, I., Van Dongen, W., and Van Onckelen, H. (1995) Quantifying phytohormones in transformed plants, in *Methods in Molecular Biology,* vol. 44: Agrobacterium Protocols. (Gartland, K. and Davey, M., eds.) Humana, Totowa, NJ, pp. 245–262.
19. Schlenk, H. and Gellerman, J. L. (1960) Esterification of fatty acids with diazomethane on a small scale. *Anal. Chem.* **32,** 1412–1414.
20. Normanly, J., Slovin, J. P., and Cohen, J. D. (1995) Rethinking auxin biosynthesis and metabolism. *Plant Physiol.* **107,** 323–329.

5

Photoacoustic and Photothermal Detection of the Plant Hormone Ethylene

Laurentius A. C. J. Voesenek, Frans J.M. Harren, Hugo S. M. de Vries, Cor A. Sikkens, Sacco te Lintel Hekkert, and Cornelis W. P. M. Blom

1. Introduction

1.1. Ethylene Detection

The hydrocarbon ethylene (C_2H_4) is a plant hormone that plays an important role in the regulation of many environmentally and developmentally induced processes, such as stress resistance, seed germination, fruit ripening, senescence, and abscission *(1)*. All tissue types and probably all cells of higher plants produce and liberate ethylene *(2)*. Many lower plants, such as liverworts, mosses, ferns, lycopods, and horse tails, also are producers of ethylene, although the biosynthetic route seems to be different *(2,3)*. Tremendous progress has been achieved during the last two decades in the biochemical and molecular characterization of the biosynthetic pathway for ethylene in higher plants *(4,5)*.

Endogenous ethylene concentrations inside plant tissues depend on the activities of certain enzymes, the rate of outward diffusion, and the rate of metabolism *(1)*. The rate-limiting reactions of ethylene biosynthesis involve the conversion of *S*-adenosylmethionine into 1-aminocyclopropane-1-carboxylic acid (ACC) catalyzed by

From: *Methods in Molecular Biology, vol. 141: Plant Hormone Protocols*
Edited by: G. A. Tucker and J. A. Roberts © Humana Press Inc., Totowa, NJ

ACC synthase and the conversion of ACC into ethylene mediated by the enzyme ACC oxidase *(6)*. Furthermore, ACC can be conjugated into 1-malonyl-aminocyclopropane-1-carboxylic acid (MACC) and 1-γ-L-glutamylaminocyclopropane-1-carboxylic acid (GACC) *(7)*. Ethylene can diffuse rapidly from plant tissues, as demonstrated in an experiment in which shoots of *Rumex palustris* were desubmerged after a submergence treatment. Of all the ethylene accumulated in the submerged shoot, 90% escaped within 1 min of desubmergence *(8)*. Furthermore, evidence increases that aerenchyma tissue, developed in many wetland plants to facilitate oxygen diffusion to root tips, is also important for outward diffusion of ethylene, thus avoiding growth inhibitory levels of this phytohormone in roots *(9)*. Higher plants can metabolize ethylene into carbon dioxide, ethylene glycol, and ethylene oxide. Usually <10% of the ethylene produced by plants is metabolized. This means that it is highly unlikely that ethylene metabolism controls endogenous ethylene concentrations in higher plants *(10)*.

The action of ethylene is not only controlled by endogenous ethylene concentrations in tissues, but also by the tissue sensitivity. It is widely assumed that molecules involved in ethylene perception and in the transduction of the signal probably control how much ethylene is required to evoke a physiological response. It has recently been demonstrated in *Lycopersicon esculentum* and *R. palustris* for various plant processes (fruit ripening, flower senescence, abscission, and flooding-induced shoot elongation) that the expression of genes coding for putative ethylene receptors positively correlate to changes in tissue sensitivity *(11–13)*.

The first chemical quantification of ethylene was performed more than 60 years ago on ripening apples *(14)*. In the years thereafter, various techniques, such as bioassays, gravimetric analyses, manometric techniques, and physicochemical colorimetric assays, were applied to quantify ethylene concentrations (*see* **ref.** ***15***). A major breakthrough in ethylene analysis was achieved in the late 1950s, when the gas chromatographic methodology was applied for the first time to ethylene *(16,17)*. For a more detailed discussion of ethylene quantification by means of gas chromatography, the columns and detectors used, and the potential sensitivity we refer to the methodological review of Bassi and Spencer *(18)*.

Despite the high sensitivity (5–10 nL/L) this gas chromatographical technique still requires accumulation of ethylene to obtain measurable quantities. For this purpose, it is common practice to incubate pieces of tissue for a few hours in small incubation vials. However, this procedure can disturb the rate of ethylene production because of wounding of tissue, disruption of transport processes, gravitropical disorientation, and changes in gas composition around the tissue *(19)*. The introduction of artifacts in measuring ethylene production rates with isolated plant tissues is elegantly demonstrated for the effect of plant-water deficit on ethylene production *(20)*. Cotton leaves, detached from the mother plant and exposed to drying conditions, showed a significant upsurge in the rate of ethylene production compared to detached control leaves. When intact cotton plants were exposed to drought no increase in ethylene production was observed *(20)*.

The only way to avoid these and other artifacts is to use larger sample chambers in which intact plants can be placed in combination with the use of continuous flow systems, to guarantee a stable gas composition. If such a setup is in line with a gas chromatograph, the concentration of ethylene in the outflowing air is usually below the detection limits of most detectors. To overcome this problem, ethylene can be concentrated on a cooled column with a high adsorption capacity for ethylene; ethylene released after heating can be measured with gas chromatography *(18,21)*. Despite the improvements, this method still has disadvantages, such as the inability to measure fast changes in ethylene production. To follow dynamic processes, it is necessary to measure ethylene directly and nearly continuously in the out flowing air of continuous flow systems *(22)*. This can be achieved if a flow through system in line with a large sampling chamber is combined with the extremely sensitive laser photoacoustic detection technique. The detection limit of this system is three orders of magnitude better than gas chromatography, i.e., 6 pL/L. Since the 1980s laser photoacoustic spectroscopy has been applied to determine ethylene production rates and concentrations in studies that focus on such processes as germination *(23,24)*, flower senescence *(25,26)*, aerenchyma formation in roots *(19)*, diffusion through aerenchymatous roots *(9)*, formation of adventitious roots *(27)*, submergence-induced shoot elongation *(8,28,29)*, and fruit ripening *(30–33)*.

In this chapter we describe the setup that allows ethylene detection at the picoliter level (laser-based photoacoustic detector) and an additional setup that makes local, nonintrusive detection of ethylene down to 0.5 nL/L possible (laser-based photothermal detector).

1.2. Photoacoustic and Photothermal Detection of Ethylene

Photoacoustic spectroscopy is based on the conversion of light energy into acoustic energy **(Fig. 1)**. Alexander Graham Bell described the photoacoustic effect for the first time more than 100 yr ago *(34)*. He showed that thin disks of very different materials, such as selenium, carbon, and rubber, emitted sound when exposed to rapidly interrupted beams of sunlight. In the years thereafter interest in the photoacoustic principle strongly declined. A renaissance of interest in this phenomenon in physics was initiated by Kreuzer *(35)*, who used powerful laser light in photoacoustic spectroscopy to measure extreme low levels of pollutants in gases. The photoacoustic effect is based on the fact that molecules absorb infrared radiation. Thus, they are excited to a higher energy level. Excited molecules fall back to their original ground state either by radiative decay (fluorescence) or nonradiative decay (collisions). In the infrared region this leads almost exclusively to nonradiative decay. De-excitation by collisional relaxation increases the kinetic energy and thus the temperature of the gas molecules around the excited molecules. If this process takes place in a constant volume it increases the pressure (Law of Boyle-Gay Lussac). A light source modulated at an audio frequency will generate pressure fluctuations at the same frequency inside the constant volume (i.e., the photoacoustic cell); this can be detected by a microphone *(36,37)*.

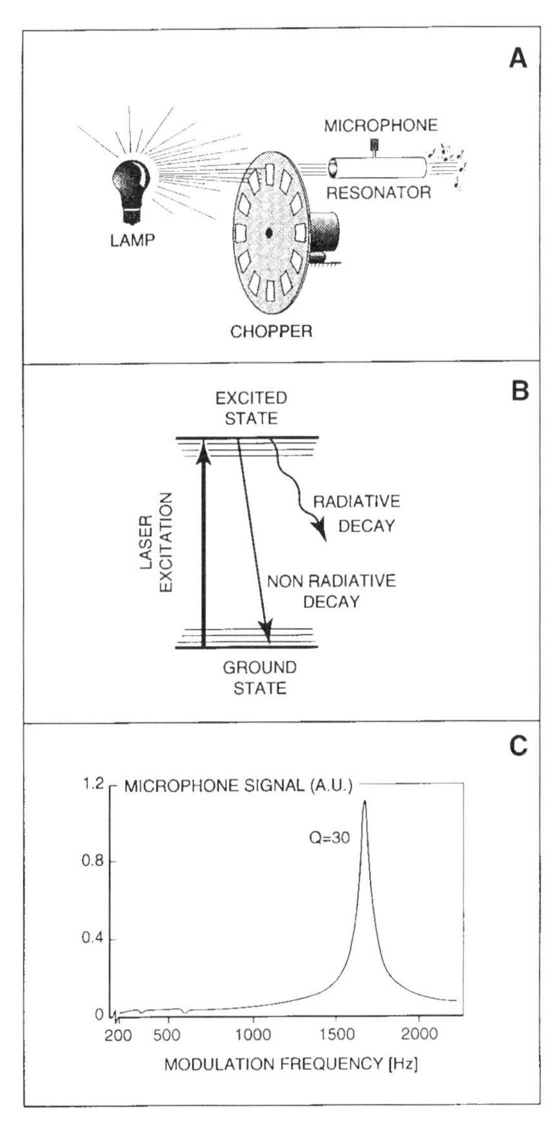

Fig. 1. **(A)** The photoacoustic effect, schematically. **(B)** A light source (sunlight, lamp, or laser) is used to excite the molecules into a higher rotational-vibrational energy level. Because of collisions the molecules return back to their original energy state, releasing heat toward the surrounding gas, resulting in a temperature and pressure increase. **(C)** The light is mechanically chopped at a frequency that corresponds to the resonance frequency of the acoustical resonator resulting in an enhancement of the photoacoustic effect.

Photothermal deflection is based on the same effect and is used to measure specific, local gas emissions at sub-ppb levels under ambient conditions. The difference with photoacoustic detection is that the temperature increase itself is detected. This increase in temperature leads to a change in refractive index in the gas. Because of the spatial gradient of the refractive index a second laser beam will be deflected. The deflection angle of the He-Ne laser beam used is an indication of the amount of molecules absorbing the CO_2 laser radiation ***(38,39)***. In the photothermal deflection setup, beams of the CO_2 and the He-Ne laser are passing the surface of an intact plant tissue at a distance of just 1.5 mm. The advantage of this method over the photoacoustic detection is that the experiments can be performed under ambient conditions (open air).

2. Materials

2.1. Photoacoustic Spectroscopy

To gain sensitivity for ethylene, high laser powers (100 W) are necessary **(Fig. 2)**. Therefore, a CO_2 laser with an extended cavity to insert a photoacoustic cell is commonly used. The setup consists of a gas discharge tube (with water jacket) to generate the laser radiation with a high voltage DC discharge through the gas filling of CO_2 (12%), nitrogen (23%), and helium (65%) (pressure 50 mbar), a grating (PTR 150 L/mm; Optometrics, Fakenham, UK) controlled by a stepmotor (Oriel Steppermike; Fairlight, Rotterdam, The Netherlands) to select CO_2 laser lines, and a chopper to modulate the CO_2 laser beam for generation of the signal inside the photoacoustic cell. The ZnSe infrared optics (Janos, Townshend, VT) consists of windows (placed at Brewster angle) to seal the discharge tube and photoacoustic cell, a positive lens (f= 35 cm), and output mirror (97% refletivity). The piezo element (Piezomechanik PST150/5/15/wc; Nema Electronics, Amsterdam, The Netherlands) adjusts the length of the laser cavity over 5 μm for optimal laser power and the power detector moni-

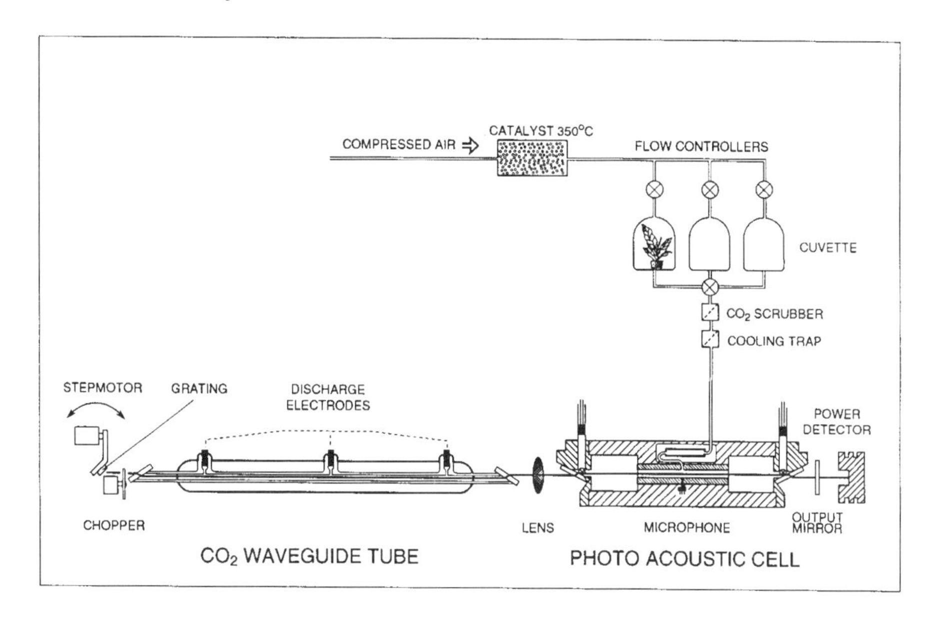

Fig. 2. The intracavity setup of the CO_2 laser with photoacoustic cell and gas flow system for measuring ethylene production from biological samples. The laser beam is generated in the discharge tube and reflected at one side by the output mirror and at the other side by the grating. The latter is also used for laser line selectivity. The gas flow system flushes the emitted ethylene from sampling cell to photoacoustic cell, passing the CO_2 scrubber and cooling trap.

tors the outcoupled laser power ***(36,37)***. Additional control and supply units consist of a computer to run the experiment, a lock-in amplifier (EG&G 5110A; Benelux B.V., Breda, The Netherlands) to filter and amplify the photoacoustic signal, a stepmotor controller, chopper controller, high voltage supply (FUG HCN350 20000; Rosenheim, Germany), piezo controller, and a laser gas mixing system including valves, mechanical flow controllers (Brooks Instrument, Veenendaal, The Netherlands), pressure indicators, and vacuum pump (Leybold Sogevac UV25; Woerden, The Netherlands) to handle the helium, nitrogen, and CO_2 needed for laser action.

A gas-flow system to flush the emitted ethylene from sampling cell to photoacoustic cell consists of a bottle of compressed air (or any other gas mixture), a catalyst containing platinized Al_2O_3 pellets (Aldrich Chemie, Bornem, Belgium) heated to 400°C to remove all traces of hydrocarbons, a pressure reduction valve, 1/16 in. tubing (De Gidts & Feldman, Almere, The Netherlands) and connectors (Amsterdam Valve & Fitting B.V., Amstelveen, The Netherlands), mechanical flow controllers (Brooks Instrument), cuvets with biological sample, a KOH-scrubber between sampling cell and a detector to remove interfering CO_2, a $CaCl_2$-scrubber to remove water, a cold-trap (110°C) to remove ethanol, and a mass flow controller (Brooks 5850S; Brooks Instrument).

Parts not assigned to a company or firm were constructed and assembled at the department of Molecular and Laser Physics, University of Nijmegen, the Netherlands.

2.2. Photothermal Spectroscopy

The intracavity CO_2 laser-based photothermal deflection setup consists of a CO_2 laser setup similar to the one described in **Subheading 2.1.** above *(39)* **(Fig. 3)**. Differences are a ZnSe prism (Laser Power Optics, San Diego, CA) driven by a galvo element (General Scanning Inc. G325D; Watertown, MA) to select laser lines and a chopper between 5 and 50 Hz. The perpendicular placed deflection frame is rigidly coupled to the CO_2 laser frame. It consists of a He-Ne laser (632.8 nm; Melles Griot 05LHP321; Zevenaar, The Netherlands), a beam splitter (Melles Griot), a multipass arrangement, and two position-sensitive quadrant detectors (Landre Intechmij B.V., Vianen, The Netherlands). Also for this detector, the parts not assigned to a company or firm were constructed and assembled at the department of Molecular and Laser Physics, University of Nijmegen, the Netherlands.

The gas-handling system can be similar to the one discussed for the photoacoustic laser setup; i.e., gas supply, water supply for laser cooling, KOH for CO_2 scrubber, assortment of Swagelok material (1/8 and 1/16 in.), and Teflon and silicon tubing material. In addition, temperature-controlled and air-ventilated conditions in the laboratory are prerequisites.

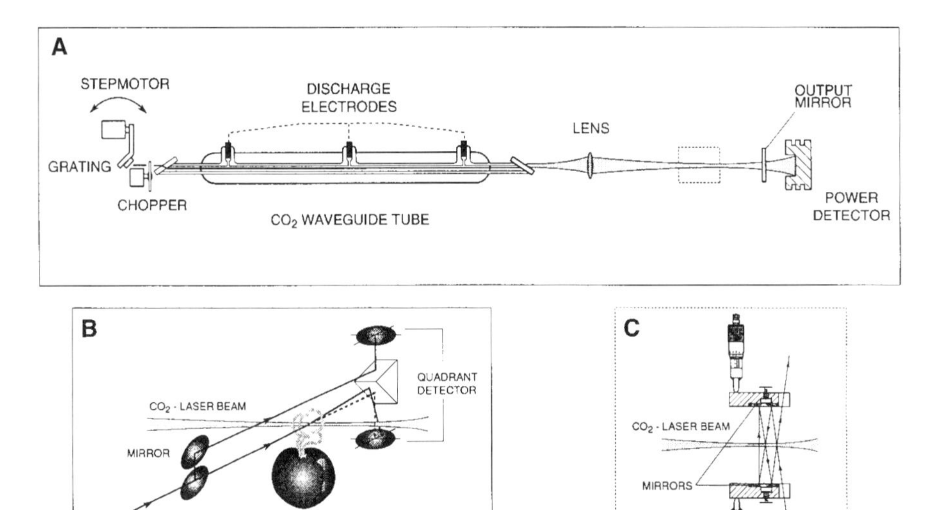

Fig. 3. The photothermal deflection setup consisting of an intracavity CO_2 laser and a dual He-Ne probe laser beam. The CO_2 laser cavity is similar as to the photoacoustic setup (**A**). The He-Ne laser beam is split in two parallel beams by a beam splitter from which one is passing the interaction region and the other is used as reference (**B**). The displacement is monitored by position-sensitive detectors. To enhance the effect the laser beams are multiple passing the interaction region in a two-mirror configuration (**C**).

3. Methods

3.1. Photoacoustic Spectroscopy

3.1.1. Alignment and Laser-Beam Generation

Each optical element of the intracavity laser setup is centered around the principal axis. A proper alignment can only be performed by an experienced laser physicist. An He-Ne laser beam can be used to align all optical elements at this principle axis, either in reflection or in transmission of the He-Ne laser beam.

Before switching on the laser discharge, proper laser gas mixture and pressure (50 mbar) should be fixed ***(36)*** and the cooling water for the discharge tube should be switched on. The total gas mass flow through the discharge is about 135 L/h at maximum laser power; the rate can be reduced (by 50%) by adjusting a valve between the gas discharge tube and the vacuum pump; then, the current must also be re-adjusted.

3.1.2. Optimization of Laser Power

Proper alignment, gas mixing, and gas pressure will generate laser power as soon as the DC discharge has been switched on. Be sure that the grating is in a position that one can expect laser radiation (e.g., at the center of the 10P branch at a wavelength of 10.6 μm). By adjusting the cavity length (i.e., the output mirror) over 5.3 μm (1/2 wavelength), the laser output power can be maximized; this can be done by a piezo element converting an applied voltage in a translation with high accuracy. The position of the photoacoustic cell inside the laser cavity can be optimized by moving the cell perpendicular to the laser beam in a vertical or horizontal direction. As soon as the outer side of the Gaussian laser beam hits the wall of the acoustic resonator the photoacoustic signal increases and the laser power drops. To assure proper alignment the cell should be filled or flushed with nitrogen gas to minimize the gas absorption signal (instead of using nitrogen, clean air passing the catalyst, scrubbers and a cold trap can also be used). The infrared optics should be kept as clean as possible and never be touched by bare hands. Cleaning should only be done with special optical tissues soaked in methanol. The grating should never be cleaned.

3.1.3. Line Selection and Optimization of Photoacoustic Signal

Laser lines can be selected by using a CO_2 infrared spectrum analyzer (Optical Engineering). However, if this analyzer is not available lines can be selected by adding traces of ethylene gas (e.g., 1 ppm) in the gas flow through the photoacoustic cell. The CO_2 laser has about 80 laser lines in the 9–11 μm region divided over four branches. Lines in the "10P" branch can be indicated using traces of ethylene, preferably in a flow of nitrogen at standard temperature

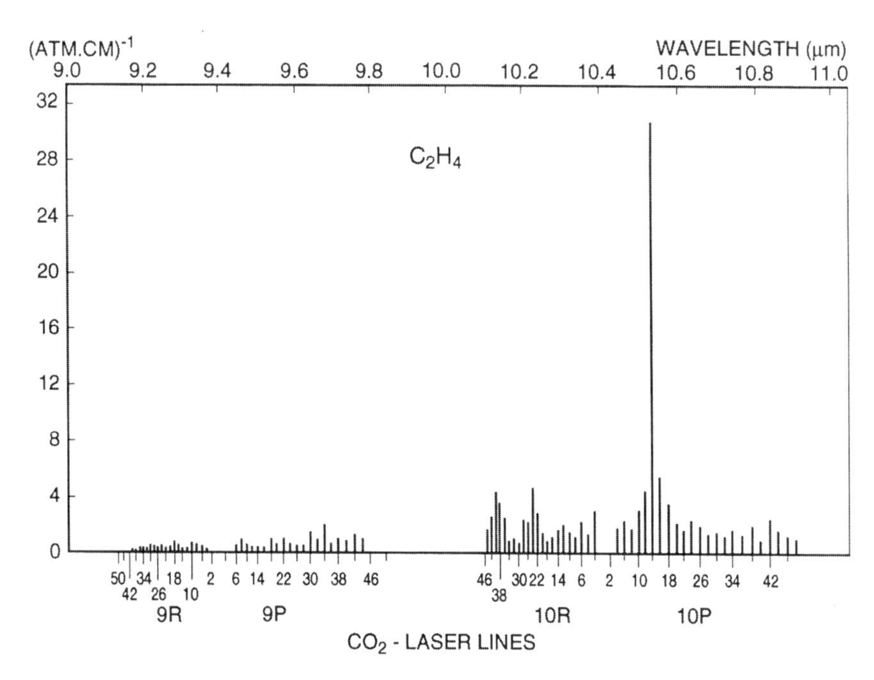

Fig. 4. The absorption pattern of ethylene at the CO_2 laser lines. The laser generates radiation at about 80 laser lines in a wavelength region between 9 and 11 µm. A distinct pattern appears at these wavelengths with the strongest absorption at the 10P14 line.

and pressure (STP); the 10P14 line demonstrates the highest absorption, i.e., six times larger than the neighboring 10P12 and 10P16 laser lines **(Fig. 4)** *(40)*.

To get the maximum photoacoustic signal the chopper has to be tuned to the resonance frequency of the photoacoustic cell. At this frequency an acoustical standing wave is generated in the open resonator **(Fig. 1)**. The speed of sound relates to the modulation frequency with the acoustical wavelength; at resonance frequency half of the acoustical wavelength is equal to the length of the resonator; the nodes of the longitudinal wave are at both ends of the open resonator. Since the speed of sound depends on temperature and gas content (oxygen, nitrogen, or CO_2 at percentage concentration) the modulation frequency must be adapted to the resonance frequency for maximal signal.

Calibration of the system is performed choosing the 10P14 laser line and, subsequently, injecting a specified volume of a certain concentration of ethylene (e.g., certified gas mixture of 1 ppm of ethylene in highly purified nitrogen) into a cuvet, which is flushed with clean air (or nitrogen) at a constant flow. The area under the obtained curve determines the relation between photoacoustic signal and the total amount of ethylene, i.e., determines the conversion factor.

3.1.4. Flow through System

Starting with compressed air (preferably not too contaminated), the air passes filters (attached to the catalyst) to remove dust and oil droplets. Inside the catalyst carbohydrates are converted to CO_2 and H_2O. The temperature of the catalyst is 400°C. Check that no water vapor is visible inside the outlet tube of the catalyst; if so, remove the outlet tube and flow air through the catalyst. The outflowing air passes a pressure-valve adjusted at 1.5 atmosphere. Thereafter, the flow is divided into different tubes. Each of these flows can be adjusted with the help of a flow controller in line. Next, the flow enters the biological sampling cuvet, of which the outlet is led to a three-way valve. One port is connected via a multi-inlet-1-outlet divider to the KOH-based CO_2 scrubber, the $CaCl_2$-based water vapor scrubber, cold trap, photoacoustic cell, and finally a mass flow controller. The other port acts as additional outlet to prevent stopping the air flow if this specific sample is not connected to the detection cell. To keep the slight overpressure inside the sampling cell, a syringe needle is placed at the second port. The KOH-based CO_2 scrubber must always be inserted into the flow through system; the $CaCl_2$ is only required if high humidity is used in the sample cuvet and to avoid freezing of the cold trap. The cold trap is used if the production of disturbing components (e.g., ethanol) can be expected. A typical temperature of the cold trap is –110°C; a lower temperature will liquify the oxygen in the air, a higher temperature will allow partial vapor pressures of interfering gases to pass the trap. Finally, the flow passes a mass flow detector, which monitors the flow rate in L/h. Depending on the ethylene production of the biological sample and the volume of the cuvet a flow rate of between 0.2 and 5 L/h is installed. The flow rate (L/h) is multiplied by the measured ethylene concentration (nL/L) to obtain a production rate (nL/h). Furthermore, production rates can be normalized to dry or fresh weight of the biological sample to allow comparisons.

3.1.5. Turning Off the System and Maintenance

After finishing an experiment the following sequence of actions safely turns off the system: Turn off the high voltage on the laser discharge tube, close the laser gas supply bottles, close any additional gas valves, and turn off the vacuum pump. Turn off the catalyst (block the flow through the catalyst approx 15 min after turning off the catalyst) and turn off all other instruments.

If any of the ZnSe windows or lenses are contaminated they have to be cleaned with methanol using special lens-cleaning tissues. Brute force must be avoided to keep from damaging the window surface. Cleaning should only be done by experienced scientists. The grating should never be cleaned or touched. The Brewster windows, especially onto the laser discharge tube, are polluted frequently on the inside surface; therefore, they must be replaced (e.g., each year at continuous operation). It is preferred to have a small (e.g., 1 L/h) flow of nitrogen gas through the gas sampling system, even when the detector is turned off, to keep it clean.

3.1.6. Measurements

The various parts of the laser-driven photoacoustic setup should be computer controlled, enabling a fully automated sampling of ethylene production rates of biological tissue. The sampling run starts by tuning the grating toward the ethylene-absorbing CO_2 laser line (10P14 laser line at 10.53 μm). Thereafter, the laser power at that laser line is maximized by adjusting the laser cavity length over half a wavelength (5.3 μm) by the piezo, which is mounted to the output mirror holder. Subsequently, the photoacoustic signal and the laser power are alternatively sampled and divided onto each other in a closed loop over (e.g.) 100 times for averaging. Samples encompassing the 95% confidence interval are used to calculate the mean ethylene signal at this 10P14 line. Hereafter, the position of the grating changes to tune the laser to a non-ethylene absorbing wavelength (i.e., 10P12 laser line at 10.51 μm wavelength). Again the laser power is optimized and the sample routine is performed. The signals on both laser lines include ethylene absorption and background signal. The latter is generated from wall and window absorption and from interfering gases. The first two signals can be distinguished quite easily because of their relative wavelength-independent absorption patterns, which should not change during the experiment. The interfering gases can cause problems in quantifying the amount of ethylene; therefore, they should be removed from the gas flow. The overall measuring sequence takes approx 1 min; ethylene concentration and background signal are calculated properly by the computer. The next step can be another photoacoustic run on the same cuvet or alternatively, a computer-controlled switch to another cuvet that is empty or contains control tissue. The latter signals should be subtracted from the experiment to extract the pure effect and cancel slow, externally induced variations (e.g., over hours or days).

3.2. Photothermal Spectroscopy

3.2.1. Alignment and Laser Beam Generation

The photothermal deflection setup consists of two parts; one frame for the CO_2 laser and another frame for the He-Ne laser. The latter frame should be rigidly fixed to the first one to avoid relative displacement of both laser beams. In principle the CO_2 laser setup is the same as the one described in **Subheading 3.1.** above except that here a sharper intracavity focus (waist 0.28 mm) is performed by the combination of a ZnSe lens (f = 75 mm) and a spherical ZnSe outcoupling mirror (inner radius of curvature R = 270 mm). The distance between lens and mirror is 350 mm, which is the available space to insert a biological sample. Care has to be taken to align the lens and outcoupling mirror properly. For optimization of the laser power, line selection, flow through system, system shutdown, and maintenance, we refer to the photoacoustic spectroscopy section.

Nearly instantaneous detection of trace gases is achieved by implementing a laser-line-selective device in the cavity of a CO_2 laser. The principle is based on rapidly switching between two laser lines, achieved by installation of a ZnSe prism, mounted on a galvo element, between the grating and discharge tube *(39)*. The prism is similar to a Brewster window of 1 in.; however, one side is inclined with respect to the other by 2°. The prism is installed at Brewster angle, minimizing power losses to 18% on the strongest laser line. The angle between the grating and incoming beam is adjusted by rotation of the prism. Using a square wave modulation for the galvo element at a repetition frequency of 10 Hz yields a stable output power at two subsequent laser lines.

3.2.2. Probe Laser Frame

The second frame consists of an He-Ne laser, a beamsplitter, a multipass mirror setup, and two quadrant detectors. All are mounted on a rigid frame to avoid disturbance by mechanical vibrations. The He-Ne laser is chosen for its small beam diameter of 0.79 mm as well as a small divergence angle of 1 mrad. The beam is divided into two beams, by means of a beam splitter, to decrease the noise caused by air turbulence (**Fig. 3**). The vertical distance between the two beams should be kept small. In this way air turbulence (diameter of vortex ranging from several meters to several millimeters) influences both beams in almost the same way. Both beams are aligned in a parallel fashion. One passes just over the heated region, i.e., at 0.7 mm from the pump beam, and will be deflected; the other, unaffected beam traverses at a distance of 10 mm over the pump beam. The distance of 0.7 mm is chosen such that the deflection will be at a maximum. The deflection is related to the gradient in refractive index. Using a chopper frequency of 20 Hz and the focused CO_2 laser configuration as described in **Subheading 3.2.1.**, the gradient is maximum at an offset of 0.7 mm from the CO_2 laser *(39)*.

The He-Ne probe laser beam is passed over the heated region several times by means of a flat mirror system. It should be noted that all beams are aligned such that each beam is passing the pump beam with an offset of 0.7 mm. The main multipass system consists of two square flat mirrors (25 × 25 mm) with highly reflective coatings, tilted with respect to each other by a small angle. The number of passes is a function of this angle and of the angle of the incident He-Ne beam with respect to the mirror. In practice, the beam divergence and the beam waist of the probe beam cannot be neglected. Thus, the distance between the mirrors should also be taken into account. In our actual setup the optimum number of passes is 31 at a tilt angle of about 1° and a mirror distance of 40 mm. The dimensions of the mirrors determine the interaction length of probe and pump beam and, thus, the spatial resolution of this setup.

3.2.3. Measurements

Within the multipass area a reference cell can be installed. In this way the detector can be calibrated by flushing a standard gas mixture through the cell (e.g., 1 ppm of ethylene in pure nitrogen). In addition biological experiments can be performed under well-defined conditions. After calibration, CO_2 laser lines can be identified by open-air experiments flushing ethylene, water, or ammonia through the interaction zone of the two laser beams. All these molecules demonstrate clear fingerprints in the CO_2 laser wavelength region ***(39)***.

To identify local trace gas emissions from biological tissue, the tissue needs to be adjusted close to the CO_2 laser beam in the interaction zone of the two beams. However, care has to be taken to avoid external influences, like heating or burning of the tissue; a typical distance of 1.5 mm between CO_2 laser beam and tissue surface is advised.

The deflection of the probe laser beam is measured with a position-sensitive detector, consisting of a diode quadrant detector with a voltage slope of 0.512 V/mm between two quadrant parts. The reference beam is detected by a separate quadrant detector. The detected signals are subtracted in a lock-in amplifier (time constant of 1 s) yielding an improvement of the sensitivity by a factor of two.

This entire experimental setup is computer controlled; in particular the grating adjustment and the piezo optimization necessary to achieve single laser line operation. The positioning of the biological sample below the crossing of the two laser beams is also adjusted by the computer. Deflection signal and intracavity CO_2 laser power are handled in the same way as for the photoacoustic setup.

4. Notes

4.1. Photoacoustic Spectroscopy

Acoustic sound from outside the photoacoustic cell will enter the microphone and will induce noise on top of the signal. Fortunately, one can diminish the surrounding noise by using a lock-in amplifier, which selectively amplifies the photoacoustic signal. The lock-in amplifier amplifies only within a certain bandwidth (e.g., 1 Hz) and in a small-phase angle around the reference frequency generated by the chopper. The reminiscent noise can be rejected by constructing the photoacoustic cell of stiff and heavy material (e.g., brass). The ultimate noise level will be the electronic noise level from the microphone itself; one should come as close as possible to it. The ratio of generated signal from the sample gas and the acoustic noise level determines the signal-to-noise level and thus the detection limit of the system.

Absorptions generated by the laser beam (window, wall) are in-phase with the modulation frequency and will be seen as a constant photoacoustic signal relatively independent of the laser line. These background absorptions should not be to high; however, they do not interfere with the sensitivity of the setup.

Another source of noise can arise from the tubing containing the flow though system and will be either flow noise from turbulences of the gas flow (lower the flow rate to check directly on the microphone signal connected to a oscilloscope) or caused by mechanical coupling with, e.g., a vacuum pump or chopper. Use coaxial cables and connectors to avoid electrical pick-up.

Other gases can interfere with the generated signal from ethylene absorption. They will not have a wavelength-independent signal that is stable over the time period of the experiment; therefore, they have to be removed from the gas flow with a cold trap.

A special case is CO_2; although the vibrational absorption performs only weakly in the 9–11 μm region the individual absorption lines are positioned exactly at the CO_2 laser lines. The fingerprint absorption will appear on these lines as a slowly changing pattern. In addition, because of the combination of the modulation frequency (>1000 Hz) and the slow collisional relaxation of the absorbed energy by CO_2, the generated photoacoustic signal will undergo a phase shift. Thus, in combination with the absorption of ethylene the overall signal will be lower and can result in negative (calculated) concentrations ***(41)***. This can be overcome by removing the CO_2 out of the gas flow between sampling cell and detection cell by a KOH scrubber.

4.2. Photothermal Spectroscopy

Photothermal deflection is very well suited to detect gases locally, next to those that tend to stick to walls and are reactive. Although less sensitive, because of the open air environment and contact-free detection, this method is advantageous as compared to photoacoustics.

Several sources of noise (at the modulation frequency of 20 Hz) have to be considered and minimized to achieve the sensitivity of 0.5 ppb for ethylene; i.e., instrumental noise, mechanical vibrations, and air turbulence. To the first group belong the electrical noise of the quadrant detector and the pointing stability of the He-Ne laser. Fluctuations in pointing stability of the CO_2 laser are negligible. The He-Ne laser, multipass system, and detector should be mounted in a rigid frame to minimize mechanical vibrations between the two laser beams. The entire setup is sustained by soft cushions to uncouple external mechanical vibrations. A low modulation frequency of 20 Hz is preferably to have a maximum gradient of refraction resulting in a maximized signal. Air turbulences generate the highest noise and are responsible for the detection limit. They can be minimized by shielding the entire setup. For the same reason the chopper is positioned between the grating and the discharge tube, as far away as possible from the detection zone to prevent air turbulence. Using a double-probe beam setup diminishes the influence of air turbulence. The detection limit of 0.5 ppb ethylene was achieved at 20 Hz, 100 W laser power, pure nitrogen as buffer gas in the detection cell, and 31 passes using the flat-mirror multipass configuration.

The temporal resolution is determined by the slowest step in the pathway from excitation to detection. Excitation is in the order of nanoseconds. For collisional decay the characteristic decay value is below 1 μs for ethylene at standard pressure and temperature. Thermal conductivity yields the optimum heat profile after a characteristic rise time of 15 ms, whereas probing of the refractive index gradient is determined by the time response of the detector (0.1 ms). Thus, the temporal resolution is determined by the thermal conductivity. However, to detect low concentrations of ethylene, changes in interfering gases need to be taken carefully into account; these, in contrast to photoacoustics, cannot easily be removed—one needs to switch between several laser lines. The switching time between two neighboring laser lines, including piezo optimization, is 15 s. This switching time has, however, been reduced to 0.1 s using the oscillating prism. Finally, the signal is fed into a lock-in amplifier; for the fast fluctuation measurements the time constant is 30 ms, for the relatively slow measurements 1 s.

The radial dimensions of the CO_2 laser beam are of importance, considering possible temperature influences on plant tissues. In experiments attention should be paid to external influences on tissues, like heating and burning. In the wings of the Gaussian beam profile a large amount of laser power is still available. In a control experiment a thermocouple placed inside a tomato just beneath the epidermis showed a temperature rise of 2°C with the tissue 1.5 mm away from the CO_2 laser beam. Compared to this the temperature rise caused by trace gas absorption is only in the order of a few mK.

The most important interfering gases for ethylene detection are CO_2 and H_2O; typical ambient concentrations are 350 ppm and 0.5–2%, respectively. For CO_2, this results in a concentration equivalent to 60 ppb of ethylene at the 10P14 laser line ***(42)***. For 1.5% water, this value corresponds to 1.4 ppb of ethylene ***(43,44)***. As mentioned before, CO_2 and H_2O have broad, unstructured absorptions patterns, which in our case is advantageous. Variations of the concentrations mainly result in a change in background signal. Subtracting the signal from the nonabsorbing, reference laser line from the signal on the absorbing laser line leaves only a small, but still significant, contribution from CO_2. As difference signal a CO_2 concentration of 350 ppm corresponds to 4.2 ppb of ethylene. Because of the low modulation frequency as compared to photoacoustics (20 instead of 1600 Hz) a phase shift does not occur. A fluctuation of 10% in absolute CO_2 concentration yields a fluctuation of 0.5 ppb for ethylene. The corresponding value for H_2O is typically smaller by a factor of 10. In experiments with plants, besides CO_2 and H_2O, other gases can be present at high concentrations, e.g., ethanol, an end product of fermentation. Fortunately, the absorption spectrum of ethanol is also broad, yielding an increase in background signal. Similarly, the role of broadband absorption of other carbohydrates, like propylene, leads mainly to an increase in background. Some of these components are cryotrapped in photoacoustic measurements; this cannot be applied in open-air photothermal deflection measurements.

4.3. General Notes

The described ethylene detectors are very sensitive for ethylene; therefore, one should always be aware of unexpected emissions. As an example, the tubing material that contains the gas sample flow may be polluted by ethylene-producing bacteria. Also UV light (and the UV part of the visible light) will dissociate the polymer tubing, generating traces of ethylene.

Both the photoacoustic and the photothermal techniques depend linearly on concentrations over the range of subppb to 100 ppm of ethylene mixed in pure nitrogen. At higher levels, the intracavity laser power is lowered substantially.

Too high laser intensities will lead to saturation effects, giving rise to lower absorption coefficients as compared to literature values ***(36)***. Attention should be paid to this aspect, because both photoacoustic and photothermal detections are using high laser powers and sharply focused laser beams.

The CO_2 laser can be replaced by a sealed-off commercial laser. It should be noted that those cannot be used intracavity, which lowers the available power and thus the sensitivity for ethylene. Implementation of both techniques in the biological field, as well as making use of an additional CO laser, allows one to study various dynamic processes. A list of gaseous compounds being detected at (sub)ppb level is presented elsewhere ***(45,46)***. Because of the complexity of the method additional information should be obtained from the authors. Experience with lasers concerning maintenance is advantageous. In addition, the described methods serve as a user facility at the University of Nijmegen (The Netherlands) and University of Bonn (Germany). To construct these laser-based detectors one should have a close collaboration between laserphysicists and biologists.

References

1. Abeles, F. B., Morgan, P. W., and Saltveit, M. E., Jr. (1992) *Ethylene in Plant Biology*. Academic Press, London, UK.
2. Osborne, D. J. (1989) The control role of ethylene in plant growth and development, in *Biochemical and Physiological Aspects of Ethylene Production in Lower and Higher Plants* (Clijsters, H., de Proft, M., Marcelle, R., and van Poucke, M., eds.) Kluwer Academic Publishers, Dordrecht, The Netherlands, pp. 1–11.
3. Osborne, D. J., Walters, J., Milborrow, B. V., Norville, A., and Stange, L. M. C. (1996) Evidence for a non-ACC ethylene biosynthesis pathway in lower plants. *Phytochemistry* **42,** 51–60.
4. Yang, S. F. and Hoffman N. E. (1984) Ethylene biosynthesis and its regulation in higher plants. *Ann. Rev. Plant Physiol.* **35,** 155–189.
5. Kende, H. (1993) Ethylene biosynthesis. *Ann. Rev. Plant Physiol. Plant Mol. Biol.* **44,** 283–307.
6. Fluhr, R. and Mattoo, A. K. (1996) Ethylene: biosynthesis and perception. *Crit. Rev. Plant Sci.* **15,** 479–523.

7. Martin, M. N., Cohen, J. D., and Saftner, R. A. (1995) A new 1-aminocyclopropane-1-carboxylic acid-conjugating activity in tomato fruit. *Plant Physiol.* **109,** 917–926.
8. Voesenek, L. A. C. J., Banga, M., Thier, R. H., Mudde, C. M., Harren, F. J. M., Barendse, G. W. M., and Blom, C. W. P. M. (1993) Submergence-induced ethylene synthesis, entrapment, and growth in two plant species with contrasting flooding resistances. *Plant Physiol.* **103,** 783–791.
9. Visser, W. J. W., Nabben, R. H. M., Blom, C. W. P. M., and Voesenek, L. A. C. J. (1997) Elongation by primary lateral roots and adventitious roots during conditions of hypoxia and high ethylene concentrations. *Plant Cell Environ.* **20,** 647–653.
10. Hall, M. A. (1991) Ethylene metabolism, in *The Plant Hormone Ethylene* (Mattoo, A. K. and Suttle, J. C., eds.) CRC, Boca Raton, FL, pp. 65–80.
11. Wilkinson, J. Q., Lanahan, M. B., Yen, H. C., Giovannoni, J. J., and Klee, H. J. (1995) An ethylene-inducible component of signal transduction encoded by *Never-ripe*. *Science* **270,** 1807–1809.
12. Payton, S., Fray, R. G., Brown, S., and Grierson, D. (1996) Ethylene receptor expression is regulated during fruit ripening, flower senescence and abscission. *Plant Mol. Biol.* **31,** 1227–1231.
13. Vriezen, W. H., Van Rijn, C. P. E., Voesenek, L. A. C. J., and Mariani, C. (1997) A homologue of the *Arabidopsis thaliana* ERS gene is actively regulated in *Rumex palustris* upon flooding. *Plant J.* **11,** 1265–1271.
14. Gane, R. (1934) Production of ethylene by some ripening fruit. *Nature* **134,** 1008.
15. Abeles, F. B. (1973) *Ethylene in Plant Biology*. Academic, New York.
16. Burg, S. P. and Stolwijk, J. A. J. (1959) A highly sensitive katharometer and its application to the measurement of ethylene and other gases of biological importance. *J. Biochem. Microbiol. Technol. Eng.* **1,** 245–259.
17. Huelin, F. E. and Kennett, B. H. (1959) Nature of the olefins produced by apples. *Nature* **184,** 996.
18. Bassi, P. K. and Spencer, M. S. (1985) Methods for quantification of ethylene produced by plants, in *Gases in Plant and Microbial Cells* (Linskens, H. F. and Jackson, J. F., eds.) Springer-Verlag, Berlin, pp. 309–321.
19. Brailsford, R. W., Voesenek, L. A. C. J., Blom, C. W. P. M., Smith, A. R., Hall, M. A., and Jackson, M. B. (1993) Enhanced ethylene

production by primary roots of *Zea mays* L. in response to sub-ambient partial pressures of oxygen. *Plant Cell Environ.* **16,** 1071–1080.
20. Morgan, P. W., He, C., De Greef, J. A., and De Proft, M. P. (1990) Does water deficit stress promote ethylene synthesis by intact plants? *Plant Physiol.* **94,** 1616–1624.
21. De Greef, J. A. and de Proft, M. (1978) Kinetic measurements of small ethylene changes in an open system designed for plant physiological studies. *Physiol. Plant.* **42,** 79–84.
22. Voesenek, L. A. C. J., Banga, M., Rijnders, J. H. G. M., Visser, E. J. W., Harren, F. J. M., Brailsford, R. W., Jackson, M. B., and Blom, C. W. P. M. (1997) Laser-driven photoacoustic spectroscopy: what we can do with it in flooding research. *Ann. Botany* **79 (suppl A),** 57–65.
23. Petruzzelli, L, Harren, F. J. M., and Reuss, J. (1994) Patterns of C_2H_4 production during germination and seedling growth of pea and wheat as indicated by a laser-driven photoacoustic system. *Environ. Exp. Bot.* **34,** 55–61.
24. Thuring, J. W. J. F., Harren, F. J. M., Nefkens, G. H. L., Reuss, J., Titulaer, G. T. M., De Vries, H. S. M., and Zwanenburg, B. (1994) Ethene production by seeds of *Striga hermonthica*-induced by germination stimulants, in *Biology and Management of Orobanche* (Pieterse, A. H., Verkleij, J. A. C., and ter Borg, S. J., eds.) Royal Tropical Institute, Amsterdam, The Netherlands, pp. 225–236.
25. Woltering, E. J., Harren, F., and Boerrigter, H. A. M (1988) Use of a laser driven photoacoustic detection system for measurements of ethylene production in *Cymbidium* flowers. *Plant Physiol.* **88,** 506–510.
26. Woltering, E. J. and Harren, F. (1989) Role of rostellum desiccation in emasculation induced phenomena in orchid flowers. *J. Exp. Bot.* **40,** 209–212.
27. Visser, E. J. W., Bögemann, G. M., Blom, C. W. P. M., and Voesenek, L. A. C. J. (1996) Ethylene accumulation in waterlogged *Rumex* plants promotes formation of adventitious roots. *J. Exp. Bot.* **47,** 403–410.
28. Voesenek, L. A. C. J., Harren, F. J. M., Bögemann, G. M., Blom, C. W. P. M., and Reuss, J. (1990) Ethylene production and petiole growth in *Rumex* plants induced by soil waterlogging: The application of a continuous flow system and a laser-driven intracavity photoacoustic detection system. *Plant Physiol.* **94,** 1071–1077.
29. Summers, J. E., Voesenek, L. A. C. J., Blom, C. W. P. M., Lewis, M. J., and Jackson, M. B. (1996) *Potamogeton pectinatus* is constitu-

tively incapable of synthesizing ethylene and lacks 1-aminocyclopropane-1-carboxylic acid oxidase. *Plant Physiol.* **111,** 901–908.

30. De Vries, H. S. M., Harren, F. J. M., Voesenek, L. A. C. J., Blom, C. W. P. M., Woltering, E. J., van der Valk, H. C. P. M., and Reuss, J. (1995) Investigation of local ethylene emission from intact cherry tomatoes by means of photothermal deflection and photoacoustic detection. *Plant Physiol.* **107,** 1371–1377.
31. De Vries, H. S. M., Harren, F. J. M., and Reuss, J. (1995) *In situ*, real-time monitoring of wound-induced ethylene in cherry tomatoes by two infrared laser-driven systems. *Post-Harvest Biol. Technol.* **6,** 275–285.
32. De Vries, H. S. M. (1996) Non-intrusive fruit and plant analysis by laser photothermal measurements of ethylene emission, in *Fruit and Nut Analyses* (Linskens, H. F., ed.) Springer Verlag, Heidelberg, Germany, pp. 1–18.
33. De Vries, H. S. M., Wasono, M. A. J., Harren, F. J. M., Woltering, E. J., van der Valk, H. C. P. M., and Reuss, J. (1996) Ethylene and CO_2 emission rates and pathways in harvested fruits investigated, *in situ*, by laser photothermal deflection and photoacoustic techniques. *Post-Harvest Biol. Technol.* **8,** 110.
34. Bell, A. G. (1880) On the production and reproduction of sound by light. *Am. J. Sci.* **20,** 305–324.
35. Kreuzer, L. B. (1971) Ultra low gas concentration infrared absorption spectroscopy. *J. Appl. Phys.* **42,** 2934–2943.
36. Harren, F. J. M., Bijnen, F. G. C., Reuss, J., Voesenek, L. A. C. J., and Blom, C. W. P. M. (1990), Sensitive intracavity photoacoustic measurements with a CO_2 waveguide laser. *Appl. Phys. B* **50,** 137–144.
37. Harren, F. J. M., Reuss, J., Woltering, E. J., and Bicanic, D. D. (1990) Photoacoustic measurements of agriculturally interesting gases and detection of C_2H_4 below the ppb level. *Appl. Spectr.* **44,** 1360–1367.
38. Jackson, W. B., Amer, N. M., Boccara, A. C., and Fournier, D. (1981) Photothermal deflection spectroscopy and detection. *Appl. Opt.* **20,** 1333–1344.
39. De Vries, H. S. M., Dam, N., van Lieshout, M. R., Sikkens, C., Harren, F. J. M., and Reuss, J. (1995) An on-line non-intrusive trace gas detector based on laser photothermal deflection. *Rev. Sci. Instr.* **66,** 4655–4664.
40. Brewer, R. J., Bruce, C. W., and Mater, J. L. (1982) Opto-acoustic spectroscopy of C_2H_4 at the 9 and 10 micrometer CO_2 laser wavelengths. *Appl. Optics* **21,** 4092–4100.

41. Rooth, R. A., Verhage, A. J. L., and Wouters, L. W. (1990) Photoacoustic measurement of ammonia in the atmosphere: influence of water vapor and carbon dioxide. *Appl. Optics* **29,** 3643–3653.
42. Rothman, L. S., Gamache, R. R., Goldman, A., Brown, L. R., Toth, R. A., Pickett, H. M., Poynter, R. L., Flaud, J. M., CamyPeret, C., Barbe, A., Husson, N., Rinsland, C. P., and Smith, A. H. (1987) The HITRAN Database, 1986 edition. *Appl. Optics* **26,** 4058–4097.
43. Ryan, J. S., Hubert, M. H., and Crane, R. A. (1983) Water vapor absorption at isotopic CO_2 laser wavelengths. *Appl. Optics* **22,** 711–717.
44. Loper, G. L., O'Neal, M. A., and Gelbwachs, J. A. (1983) Water vapor continuum CO_2 laser absorption spectra between 27°C and 10°C. *Appl. Optics* **22,** 3701–3710.
45. Bijnen, F. G. C., De Vries, H. S. M., Harren, F. J. M., and Reuss, J. (1994) Cockroaches and tomatoes investigated by laser photoacoustics. *J. Phys.* IV **C7,** 435–443.
46. Harren, F., Reuss, J. (1997) Photoacoustic spectroscopy in, *Encyclopedia of Applied Physics* (Trigg, G. L., ed.) VCH Publishers, Weinheim, Germany, pp. 413–435.

6

Analysis of Gibberellins

Stephen J. Croker and Peter Hedden

1. Introduction

Gibberellins (GAs) are involved in a wide range of plant developmental processes. The biosynthetic pathway is discussed in Chapter 1 but it should be noted here that of all the plant hormones the GAs represent perhaps the most diverse group, with currently 125 different structures known. This diversity compounds the analysis of GAs in that not only must extraction ensure quantitative recovery of all the types, but these need to be resolved following partial purification. Moritz and Olsen *(1)* have described the analysis of GAs from milligram amounts of plant tissues.

2. Materials

2.1. Qualitative Analysis

2.1.1. Extraction

1. Methanol:water (4:1 v/v).
2. 833 Bq Each of the following tritiated GA standards, GA_{19}, GA_{20}, GA_1, GA_4, and GA_9.
3. Methanol.
4. 2 *M* HCl.
5. Ethyl acetate.

From: *Methods in Molecular Biology, vol. 141: Plant Hormone Protocols*
Edited by: G. A. Tucker and J. A. Roberts © Humana Press Inc., Totowa, NJ

6. 5% (w/v) sodium bicarbonate.
7. 1 *M* KOH.
8. PVP (Polyclar AT, sieved to remove fine (<100 mesh) particles) column (3 mL bed vol) pre-equilibrated with 0.1 *M* KH_2P0_4 (pH 3).
9. QAE Sephadex A25 (Pharmacia [Uppsala, Sweden]) anion exchange column (5 mL bed vol) pre-equilibrated with sodium formate (0.5 M).
10. 0.2 *M* formic acid.
11. C_{18} SepPak cartridge (Waters Associates [Watford, Hertfordshire, UK]).

2.1.2. Derivatization and Resolution of GAs.

1. Methanol.
2. 2 m*M* Acetic acid.
3. Hypersil 5 μm octadecylsiloxane (ODS) high-performance liquid chromatography (HPLC) column (4.9 mm id × 250 mm).
4. Ethereal diazomethane.
5. *N*-Methyl-*N*-trimethylsilyltrifluoroacetamide (MSTFA).
6. Ethyl acetate.
7. Aminopropyl cartridge (100 mg).

2.2. Quantitative Analysis

As above plus:

1 $[17\text{-}^2H_2]$GAs, the appropriate standard for each GA to be quantified.
2 A gas chromatography-mass spectrometry (GC-MS) system capable of selected ion monitoring (SIM) or selected reaction monitoring (SRM). An example of this is a Hewlett-Packard 5890 gas chromatograph coupled to an HP5970 mass selective detector.
3 A fused silica WCOT capillary column (25 m × 0.22 mm × 0.25 μm film thickness) coated with either BP1 or SE52/54 or equivalent.

2.3. Analysis by GC-MS with Selected Reaction Monitoring Using a Simplified Purification Procedure

1. QAE Sephadex A25 (Pharmacia) anion-exchange column (5 mL bed vol) pre-equilibrated with sodium formate (0.5 *M*).
2. 0.2 *M* formic acid.

3. C_{18} SepPak cartridge (Waters Associates).
4. Methanol.
5. 2 m*M* acetic acid.
6. Hypersil 5 µm ODS HPLC column (4.9 mm id × 250 mm).
7. Ethereal diazomethane.
8. MSTFA.
9. Ethyl acetate.
10. Aminopropyl cartridge (100 mg).
11. An external source ion trap mass spectrometer (Finnigan GCQ [Hemel Hempstead, UK]) or equivalent.
12. CH_2Cl_2. (10–20 µL).
13. A fused silica WCOT BPX5 capillary column (25 m × 0.22 mm × 0.25 mm film thickness) (Scientific Glass Engineering) or equivalent.

3. Methods

3.1. Qualitative Analysis

3.1.1. Extraction

1. Plant tissues are frozen immediately in liquid N_2 and stored at –80°C or freeze-dried and stored at –20°C.
2. Samples (10–50 g fresh weight) are homogenized in cold (4°C) methanol:water (4:1 v/v) (>10 mL solvent/g fresh weight of tissue). Small amounts (833 Bq each) of the following tritiated GA standards, GA_{19}, GA_{20}, GA_1, GA_4, and GA_9, are added to the homogenate and the extract stirred overnight in the cold (4°C).
3. After filtration, the residue is re-extracted with methanol (>10 mL solvent/g fresh weight of tissue) for 4 h and refiltered. Methanol is removed from the combined filtrates under reduced pressure at 35°C, and the aqueous residue is adjusted to pH 2.5 (2 *M* HCl), and partitioned against ethyl acetate (4× equal vols).
4. The combined organic phases are partitioned against 5% (w/v) sodium bicarbonate (3 × 1/5 vol). The aqueous phases are acidified to pH 3.0 (2 *M* HCl) and partitioned against ethyl acetate (4× equal vol), which is then reduced to dryness *in vacuo* at 35°C.
5. The extract is dissolved in 5 mL water, adjusted to pH 8.0 (1 *M* KOH), and loaded onto a PVP (Polyclar AT, sieved to remove fine [<100 mesh] particles) column (3 mL bed vol) pre-equilibrated with

0.1 *M* KH_2PO_4 (pH 3.0). After loading, the column is washed with water (2 × 5 mL) at pH 8.

6. The eluate is combined and loaded onto a QAE Sephadex A25 (Pharmacia) anion-exchange column (5 mL bed vol) pre-equilibrated with sodium formate (0.5 *M*) then washed with formic acid (0.2 *M*) and water (pH 8.0). After loading, the column is washed with water (pH 8.0) (15 mL) and GAs are eluted with 0.2 *M* formic acid (20 mL).
7. The eluate is then applied directly to a pre-equilibrated C_{18} SepPak cartridge (Waters Associates). After washing with water (pH 3.0) (5 mL), GAs are eluted with methanol:water (4:1 v/v) (5 mL), and are then evaporated to dryness *in vacuo*.

3.1.2. Derivatization and Resolution of GAs

1. Samples are dissolved in 80 μL methanol, made up to 400 μL with 2 m*M* acetic acid.
2. GAs are resolved by reverse-phase HPLC using a 4.9 mm id × 250 mm column containing Hypersil 5 μm ODS. Samples are injected onto the column using a Rheodyne 7125 valve fitted with a 500 μL loop and eluted using a linear gradient of increasing methanol in 2 m*M* acetic acid (20 to 100% methanol over 40 min) at a flow rate of 1 mL/min. Forty 1-mL fractions are collected.
3. Aliquots (50 μL) are removed for scintillation counting to locate the tritiated standards. Fractions, based on the location of these GAs, are combined and taken to dryness *in vacuo*.
4. The dried fractions are methylated with excess ethereal diazomethane, transferred to glass ampules, and trimethylsilylated with MSTFA (5 mL) at 90°C for 30 min.
5. If necessary, further purification can be achieved by dissolving the methylated fraction in ethyl acetate (1 mL), partitioning against water (1 mL), and then passing the ethyl acetate phase through a small (100 mg) aminopropyl cartridge that has been pre-equilibrated with ethyl acetate.
6. The aqueous phase is partitioned once more against ethyl acetate (1 mL), and the organic phase passed though the aminopropyl cartridge.
7. The combined ethyl acetate phases are taken to dryness and then transferred to ampules and trimethylsilylated as described in **Subheading 3.1.2.4.**
8. Derivatized samples are analyzed by combined GC-MS in full-scan mode. Samples (1 μL) are coinjected with Parafilm in hexane to

determine KRI values and the mass spectra and KRIs are compared with those of standards or with published data *(2)*.

3.2. Quantitative Analysis

1. Samples (5–20 g fresh weight) are extracted as described in **Subheading 3.1.1.2.** and in addition to the tritiated standards, $[17\text{-}^2H_2]$GAs are added as internal standards. The appropriate standard is added for each GA to be quantified; the amounts of standards added should approximate the anticipated amount of endogenous analyte determined from preliminary experiments.
2. After purification as described in **Subheading 3.1.1.**, but omitting the PVP step, samples are analyzed by GC-MS with SIM using, for example, a Hewlett-Packard 5890 gas chromatograph coupled to an HP5970 mass selective detector.
3. Typically, samples (1–3 mL) are injected into a fused silica WCOT capillary column (25 mm × 0.22 mm × 0.25 µm film thickness) coated with either BP-1 or SE52/54 at an oven temperature of 60°C. After 0.5 min, the splitter (50:1) is opened and 1 min later the temperature is increased at 20°C/min to 240°C and then at 4°C/min to 300°C. The He inlet pressure is 0.09 MPa and the injector, interface and MS source temperatures are 220, 270, and 200°C, respectively.
4. Characteristic ions are monitored with dwell times of 50 ms (for example, GA_8, *m/z* 594,448; $[17^2H_2]GA_8$, *m/z* 596, 450; GA_1, *m/z* 506, 448; $[17^2H_2]GA_1$ *m/z* 508, 450; GA_3, *m/z* 504; $[17^2H_2]GA_3$, *m/z* 506; GA_{20}, *m/z* 418, 375; $[17^2H_2]GA_{20}$, *m/z* 420, 377; GA_{19}, *m/z* 434, 402; $[17^2H_2]GA_{19}$, *m/z* 436, 404). The concentration of GAs in the original extracts are determined from previously established calibration curves of the peak area ratios for unlabeled and deuterated GAs plotted against varying molar ratios of the two compounds *(3)*.

3.3. Analysis by GC-MS with SRM Using a Simplified Purification Procedure

This method is used in conjunction with SRM for quantitative analysis using 0.5–5 g fresh weight of tissue. The method is the same as that described in **Subheading 3.1.1.** except that the solvent partitioning and PVP steps are omitted.

1. The extract is taken to dryness, then dissolved in water (5 mL), adjusted to pH 8.0 (1 *M* KOH), and loaded directly onto the QAE

Table 1
Characteristic Parent and Product Ions for Representative GAs and Internal Standards

Gibberellin	Excitation voltage	Parent ion, *m/z*	Product ion mass range, *m/z*	Product ion, *m/z*
GA_1	0.8	506	350–506	448
$[17\text{-}^2H_2]GA_1$	0.8	508	350–508	450
GA_4	0.7	284	150–284	224
$[17\text{-}^2H_2]GA_4$	0.7	286	150–286	226
GA_8	0.9	594	430–594	448
$[17\text{-}^2H_2]GA_8$	0.9	596	430–596	450
GA_{15}	0.8	284	180–284	239
$[17\text{-}^2H_2]GA_{15}$	0.8	286	180–286	241
GA_{19}	0.8	434	250–434	374
$[17\text{-}^2H_2]GA_{19}$	0.8	436	250–436	376
GA_{20}	0.8	418	300–418	375
$[17\text{-}^2H_2]GA_{20}$	0.8	420	300–420	377
GA_{24}	0.7	314	200–314	286
$[17\text{-}^2H_2]GA_{24}$	0.7	316	200–316	288
GA_{29}	0.9	506	350–506	477
$[17\text{-}^2H_2]GA_{29}$	0.9	508	350–508	479
GA_{34}	0.8	506	350–506	416
$[17\text{-}^2H_2]GA_{34}$	0.8	508	350–508	418
GA_{44}	0.8	432	200–432	207
$[17\text{-}^2H_2]GA_{44}$	0.8	434	200–434	209
GA_{53}	0.8	448	350–448	388
$[17\text{-}^2H_2]GA_{53}$	0.8	450	350–450	390

Sephadex A-25 (Pharmacia) anion-exchange column. Thereafter, the procedure is the same as above.

2. For GC-MS analysis using SRM, an external source ion trap mass spectrometer (Finnigan GCQ) is used.
3. Derivatized samples are diluted with CH_2Cl_2 (10–20 µL) and injected (1 µL) into a fused silica WCOT BPX5 capillary column (25 m × 0.22 mm × 0.25 mm film thickness) (Scientific Glass Engineering) at an oven temperature of 60°C. After 1 min the splitter 50:1 is opened and the temperature is increased at 20°C/min to 220°C

and then at 4°C/min to 300°C. The He flow is controlled at a constant linear velocity of 40 cm/s. The injector, interface, and MS source temperatures are 220, 270, and 200°C, respectively. The mass spectrometer is operated in dual SRM mode.

4. Characteristic parent ions are selected and monitored with an isolation wave-form notch of 1 amu. Full product ion scans at 0.5 s/scan are obtained for all GAs analyzed. The concentration of GAs in the extracts is determined from previously established calibration curves of peak area ratios of the product ion for unlabeled and deuterated GAs plotted against varying molar ratios of the two compounds. Parent and product ions for representative GAs and internal standards are listed in **Table 1**.

4. Notes

1. During extraction, for many plant species the PVP purification step can be omitted.
2. If the samples are evidently very dirty, further purification can be achieved prior to trimethylation by dissolving the methylated samples in ethyl acetate (1 mL), partitioning against an equal volume of water and passing through an aminopropyl column as described in **Subheadings 2.5.–2.6.**

References

1. Moritz, T. and Olsen, J. E. (1995) Comparison between high resolution selected ion monitoring and four-sector tandem mass spectrometry in quantitative analysis of gibberellins in milligram amounts of plant tissue. *Anal. Chem.* **67,** 1711–1716.
2. Gaskin, P. and MacMillan, J. (1992) *GC-MS of the Gibberellins and Related Compounds: Methodology and a Library of Spectra.* Cantock's Enterprises, Bristol, UK.
3. Croker, S. J., Gaskin, P., Hedden, P., MacMillan, J., and MacNeil, K. A. G. (1994) Quantitative analysis of gibberellins by isotope dilution mass spectrometry: a comparison of the use of calibration curves, an isotope dilution fit program and arithmetical correction of isotope ratios. *Phytochem. Anal.* **5,** 74–80.

7

Cytokinins

Extraction, Separation, and Analysis

Paula E. Jameson, Huaibi Zhang, and David H. Lewis

1. Introduction

Analysis of cytokinins in plant tissues has historically been laborious and time consuming because of the extremely low levels and diverse molecular structures of the cytokinins. The initial anticipation that cytokinins could be quantified by radioimmunoassay of crude plant extracts *(1,2)* was, unfortunately, not justified. Detailed analysis requires the sample to be well purified and the individual cytokinins to be separated, so that not only is interference in the assay avoided, but also problems associated with differential crossreactivity of individual cytokinins with the antibodies in the immunoassay *(3)*. The (sometimes subtle) differences in the chemical properties of the cytokinin free bases, ribosides, nucleotides, and the *O*- and *N*-glucosides enables the ready separation and quantification of the majority of the cytokinins.

We describe in this chapter an efficient multiple-sample processing capability in which the analysis has been streamlined in order to deal with a large number of samples. Initial extraction of the plant material is followed by the removal of phenolics and other impurities using a polyvinylpolypyrrolidone powder (PVPP) column. To

From: *Methods in Molecular Biology, vol. 141: Plant Hormone Protocols*
Edited by: G. A. Tucker and J. A. Roberts © Humana Press Inc., Totowa, NJ

purify the sample and simultaneously achieve the separation of the cytokinin nucleotides from the free base, riboside, and glucoside forms, the sample is passed through columns containing PVPP, DE52, and C_{18}, which are linked in series. Following purification, bulk separation of cytokinin glucosides from the bases and ribosides is achieved by using normal phase high-performance liquid chromatography (HPLC) with an amine column. Separation of individual cytokinins then occurs during reverse-phase C_{18} HPLC. Quantification of the cytokinins in the resulting fractions is performed by radioimmunoassay (RIA). The entire procedure is outlined in **Fig. 1**.

Most anticytokinin antibodies do not recognize the *O*-glucosides, but the aglycones of these sugar conjugates are readily detected and can be released with treatment by β-glucosidase. We have found that the cytokinin nucleotides are most readily separated and quantified once hydrolysed to their respective ribosides by alkaline phosphatase. The procedure outlined has not been optimized for the detection of the 7-glucosides, the xylose or *O*-acetyl conjugates, or lupinic acid. References to these conjugates may be found in the review by Jameson ***(4)***.

1.1. Extraction

The initial extraction of the cytokinins should be carried out carefully to prevent enzymatic and chemical changes and to increase extraction efficiency ***(5)***. Homogenizing plant tissue in methanol has been a traditional method and is still widely used ***(6–8)***. However, cytokinin nucleotides may be hydrolysed in methanolic extracts, because phosphatases can function in such solutions ***(9,10)***. To eliminate the problem associated with phosphatases, some authors routinely use Bieleski's solution (methanol:chloroform:acetic acid:water; ***9***) to extract cytokinins from plant tissues ***(11–13)***. A modified version has also been used ***(14,15)***.

Fig. 1. Procedure used for extraction, purification, separation, and quantification of different molecular configurations of cytokinins. □ indicates column used; ⁚ ⁚ indicates columns are linked in series.

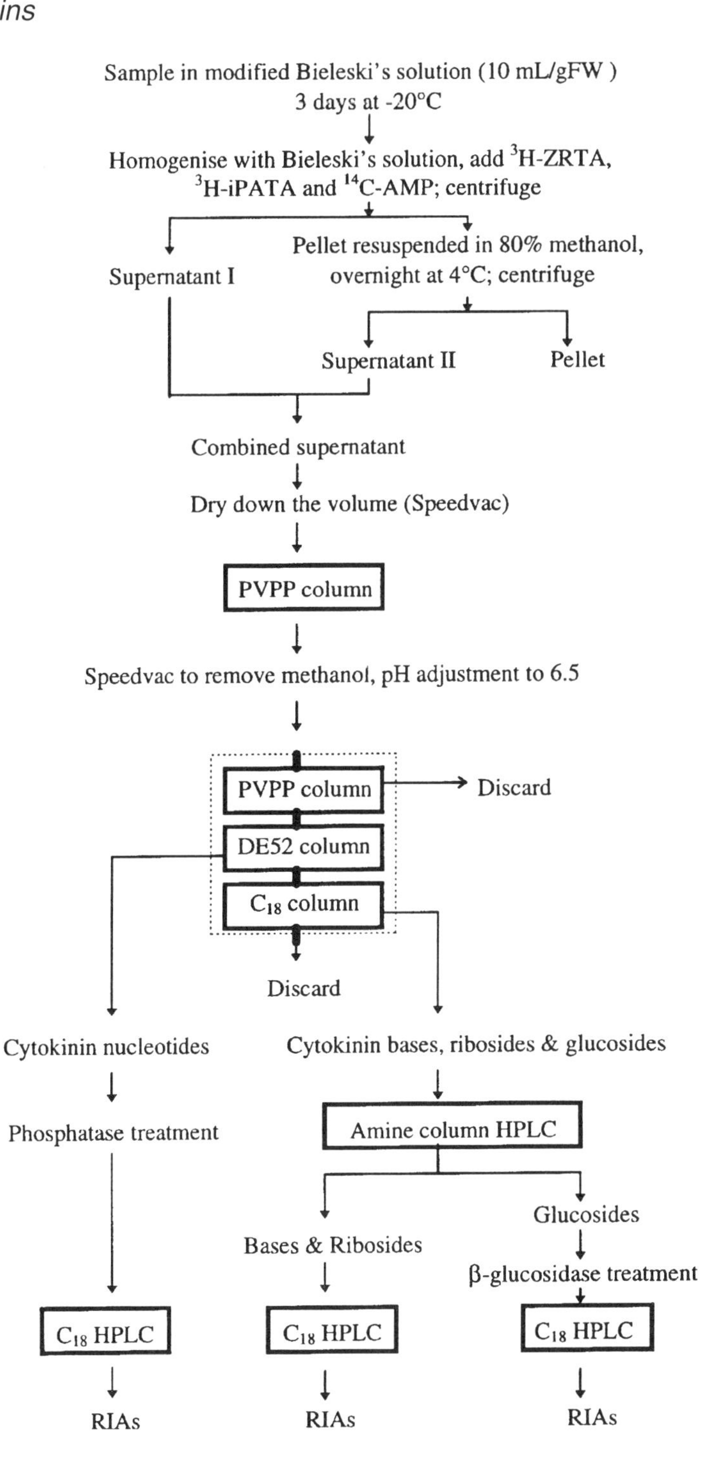

Sample in modified Bieleski's solution (10 mL/gFW)
3 days at -20°C
Homogenise with Bieleski's solution, add ^{3}H-ZRTA, ^{3}H-iPATA and ^{14}C-AMP; centrifuge
Supernatant I
Pellet resuspended in 80% methanol, overnight at 4°C; centrifuge
Supernatant II
Pellet
Combined supernatant
Dry down the volume (Speedvac)
PVPP column
Speedvac to remove methanol, pH adjustment to 6.5
PVPP column
Discard
DE52 column
C_{18} column
Discard
Cytokinin nucleotides
Cytokinin bases, ribosides & glucosides
Phosphatase treatment
Amine column HPLC
Glucosides
Bases & Ribosides
β-glucosidase treatment
C_{18} HPLC
C_{18} HPLC
C_{18} HPLC
RIAs
RIAs
RIAs

1.2. Sample Purification

Sample purification is generally achieved by the use of a range of ion-exchange and solid-phase extraction columns. Insoluble PVPP may be used for sample purification ***(16,17)***, solid-phase extraction ***(18)***, and cytokinin separation ***(19)***. Cytokinins are readily eluted from PVPP columns at low pH (3.5), whereas phenolics are retained with high affinity ***(19)***. Methanol is a very efficient solvent with which to elute cytokinins from PVPP columns ***(20)***.

During sample purification it is relatively easy to separate the nucleotide cytokinins from the other forms because the additional phosphate moiety aquires negative charges over a range of pH. The separation is usually achieved by passage of sample through a cellulose phosphate column ***(12,21,22)***. However, because cellulose phosphate is a relatively weak cation-exchange material at low pH (3.0), a large bed volume is required. As a consequence, a large volume of eluent is required to initially elute the nucleotide fraction (acid wash) and to subsequently elute the remaining cytokinins (basic wash). Therefore, a considerable volume of aqueous eluent has to be reduced to a few milliliters (usually under reduced pressure by rotary evaporation). Furthermore, use of solid-phase extraction by direct attachment of a C_{18} column to the cellulose phosphate column is not practical because acidic (pH 3.0) and basic (pH 9.0–10.0) washes are employed as eluents. Under these pH extremes the nucleotides and some of the ionized cytokinins are not retained by the C_{18} ***(23)***.

DEAE cellulose (DE52 anion exchange) has been used in sample purification as a pre-column to an immunoaffinity column ***(24)***, to trap nucleotides, including those not retained by cellulose phosphate ***(11,22)***, and to remove impurities. At neutral pH, the free bases, ribosides, and glucosides are not retained by the DEAE cellulose but the nucleotides are. In addition, at a neutral pH, a column of C_{18} connected directly to the DE52 column, will trap the cytokinins passing through the DE52 column. Since their first use in cytokinin extraction ***(25)***, open columns of C_{18} have been widely used and the linkage of DE52 and C_{18} columns has been previously reported ***(26)***.

1.3. High Performance Liquid Chromatography

Bulk separation of cytokinin glucosides from bases and ribosides has to be achieved if the storage *O*-glucosides and inactive *N*-glucosides are of interest. Thin layer chromatography (TLC) has been shown to be very useful in the separation of glucosides from the bases and ribosides *(27)*, but this technique has not been widely adopted. We have developed a normal-phase HPLC procedure to provide a bulk separation of free bases, ribosides, and *O*- and *N*-glucosides *(28)*.

Subsequently, we employ reverse-phase HPLC of the prepurified and partially fractionated sample. Reverse-phase HPLC has become an essential step prior to RIA *(24)* because individual cytokinins can not only be efficiently separated from each other but also separated from impurities impacting on the RIA *(29,30)*.

1.4. Radioimmunoassay

Since RIA was introduced into cytokinin analysis *(2)*, considerable progress has been made in speed, sensitivity, and sample handling capability *(22,24)*. However, the multiple and sequential additions of buffer, labeled analog of cytokinin, and antibody take up a large proportion of the time and effort of the RIA operation. We outline in **Subheading 3.5.**, a more streamlined protocol. Finally, the RIA should be validated because the accuracy of the assay will be influenced if interfering substances are present. For a detailed discussion on immunoassay validation refer to Pengelly *(31)* and Banowetz *(29)*. Briefly, sample validation can be carried out using dilution curves, but optimal validation requires quantification by combined gas chromatography-mass spectroscopy or by liquid chromatography-mass spectrometry using stable isotopes as internal standards. Deuterated internal standards for cytokinins are available from Apex (Honiton, UK). Although careful RIA analyses provide excellent quantification of cytokinins in multiple samples, the relatively new technique of electrospray tandem mass spectrometry must be seen as the method of choice for both identification and quantification of individual cytokinins *(32)*. Those who can access new models should find this a highly sensitive technique. With the cytokinins applied to the MS in the liquid phase, the necessity for derivatization is bypassed.

2. Materials

Unless specified otherwise, all materials are Analar grade or its equivalent, and all solvents of HPLC grade purity. MilliQ rated water is used. All procedures are performed in polypropylene containers or silanized glassware (*see* **Note 1**) to minimize losses of cytokinin. The procedure outlined has been optimized for the extraction of 2–20 g FW vegetative tissue.

2.1. Sample Extraction

1. Modified Bieleski's solution: methanol:H_2O:acetic acid 70:30:3 (v/v/v); 10 mL/g FW; pre-chilled to –20°C.
2. ^{3}H-ZR-trialcohol, ^{3}H-iPA-trialcohol, ^{14}C-AMP as internal standards (*see* **Note 2**)
3. 80% Methanol.

2.2. Sample Purification

1. Polyvinylpolypyrrolidone powder (Sigma Chemical Co. [St. Louis, MO] P-6755).
2. Methanol:0.5 *N* acetic acid, 80:20 (v/v).
3. DEAE cellulose (Whatman Biosystems Ltd [Kent, England] DE52).
4. 10 m*M* ammonium acetate buffer (pH 6.5).
5. 1 *N* acetic acid (pH 2.5).
6. Octadecyl silica (C_{18} Bondesil [Varian Institute Group, Walnut Creek, CA]).
7. Methanol.

2.3. Enzyme Treatments

1. Alkaline phosphatase treatment for nucleotide degradation: 0.1 *M* ethanolamine pH 9.5 (using concentrated HCl to adjust pH); 40 m*M* $MgCl_2$; alkaline phosphatase (Sigma EC3.1.3.1). Phosphatase reaction solution: 5 mL 0.1 *M* ethanolamine (pH 9.5); 0.1 mL 40 m*M* $MgCl_2$; 20 µL alkaline phosphatase. Dry sample to a small volume (approx 500 µL) then add 5 mL of phosphatase reaction solution.

2. β-Glucosidase treatment to degrade cytokinin *O*-glucosides: 50 m*M* sodium acetate buffer, pH 5.4; β-glucosidase (from sweet almonds, Boehringer Mannheim [GmbH, Germany]). β-Glucosidase reaction solution: 3 mg enzyme in 5 mL sodium acetate buffer, pH 5.4. Dry down sample to approx 500 μL. Dissolve in 5 mL buffer then add 5 mL β-glucosidase reaction solution to give a final enzyme concentration of 0.3 mg/mL.

2.4. High Performance Liquid Chromatography

1. Amine column (Alphasil $5NH_2$; 250 × 4.6 mm) (HPLC Technology Ltd, [Cheshire, England]).
2. Reverse phase C_{18} column (Beckman Instruments [Fullerton, CA] Ultrasphere 5 μm, 250 × 4.6 mm).
3. 40 m*M* acetic acid.
4. Triethylamine (stored under nitrogen).
5. Methanol and acetonitrile.

2.5. Radioimmunoassay

1. 50 m*M* Sodium phosphate buffer with 0.14 *M* NaCl, pH 7.2, 0.1% (w/v) gelatin (BDM Laboratory Supplies [Poole, England]), 0.01% ovalbumin (Sigma Grade V).
2. ^{3}H-ZR-trialcohol or ^{3}H-iPA-trialcohol.
3. Monoclonal antibodies crossreactive with hydroxylated, nonhydroxylated, and N-linked (i.e., 9-glucoside) cytokinins (*see* **Note 3**).
4. 90% Saturated ammonium sulfate (Rectapur grade Rhone-Poulenc Ltd, [Manchester, England]) adjusted to pH 7.0 with concentrated ammonia.
5. Scintillation cocktail (Optiphase HiSafe 2; Wallac [Milton-Keynes, England]).

2.6. Cytokinin Standards

The cytokinin standards zeatin (Z), dihydrozeatin (DZ), cis-zeatin (cZ), zeatin riboside (ZR), dihydrozeatin riboside (DZR), kinetin, isopentenyladenine (iP), and isopentenyladenosine (iPA) were obtained from Sigma. Zeatin-9-glucoside (Z9G), zeatin-*O*-glucoside (ZOG), zeatin riboside-*O*-glucoside (ZROG), dihydrozeatin-

9-glucoside (DZ9G), dihydrozeatin-*O*-glucoside (DZOG), dihydrozeatin riboside-*O*-glucoside (DZROG) and isopentenyladenine-9-glucoside (iP9G) were obtained from Apex Chemicals (Honiton, Devon, England).

3. Methods

3.1. Sample Extraction

1. As soon as possible after harvesting, weigh the plant material and place in modified Bieleski's solution, prechilled at –20°C ***(13)*** (*see* **Notes 4** and **5**).
2. The plant material should remain in the Bieleski's solution for at least 3 d at –20°C to inactivate enzymes before material is ground. After 3 d, transfer the plant material to a mortar and pestle and homogenize it using the extraction solvent.
3. Internal standards should be added into the homogenate: Add approx 30,000 CPM of ^{3}H-ZR-trialcohol, ^{3}H-iPA-trialcohol, and ^{14}C-AMP internal standards to the sample to enable an estimate to be made of losses of cytokinin during the analysis (but *see* **Notes 2** and **7**).
4. Centrifuge the homogenate at 6500*g* for 30 min at 4°C. Decant the resulting supernatant into a 50-mL Falcon tube and store at 4°C (*see* **Note 6**).
5. Resuspend the pellet in 80% methanol (5 mL/g FW), leave overnight at 4°C, and then centrifuge the sample again. Remove the second supernatant and combine it with the first. The combined supernatants are then reduced *in vacuo* (Savant Speed-Vac) to 2 mL. Samples may be stored at -20°C at this stage, prior to loading onto a preconditioned PVPP column.

3.2. Polyvinylpolypyrrolidone Column Chromatography

3.2.1. Column Preconditioning

Slurry the dry PVPP powder with acidified methanol (methanol:0.5 *N* HAc; 80:20 [v/v]). Allow the slurry to settle before carefully decanting fine particles. Place two disks of Whatman No. 4 filter paper in the bottom of a 30 mL syringe barrel to hold the PVPP in place. Pour the slurry into the column and allow it to settle under gravity to a final bed volume (BV) of 15–20 mL (allowing 2 mL

BV/g FW). Wash the packed column with further acidified methanol (2 BV) before applying the sample to the top of the column. The preconditioned column should not be allowed to dry.

3.2.2. Sample Elution and pH Adjustment

Samples are eluted by adding 3 BV of acidified methanol. Evaporate the eluent *in vacuo* to the aqueous phase (about 5 mL), adjust pH to 6.5 with 0.1 *N* NaOH, and continue evaporation to obtain a volume of 2 mL in preparation for application to the PVPP-DE52-C_{18} column complex. (**Note:** *see* **Note 8** regarding pH adjustment).

3.3. PVPP-DE52-C_{18} Column Complex Chromatography

This column complex is used both for further purification and for the separation of cytokinin nucleotides from the free bases and glucosides. The columns are individually packed and then linked in series as shown in **Fig 1**.

3.3.1. PVPP Preconditioning

Prepare the PVPP column by initially slurrying PVPP powder in 10 m*M* ammonium acetate buffer (pH 6.5). Pack the slurry into a 30-mL syringe barrel to a final BV of 5 mL (*see* **Note 9**).

3.3.2. DE52 Preconditioning

Prepare the anion exchange cellulose (DE52, Whatman) according to the manufacturer's instructions with the final equilibration in 10 m*M* ammonium acetate buffer (pH 6.5). Allow 2 mL BV/g FW (*see* **Note 10**). Pack the DE52 under gravity into a 30-mL syringe barrel to a final BV of 20 mL. Wash the column with an additional 3 BV of 10 m*M* ammonium acetate buffer (pH 6.5) before linking to the PVPP and C_{18} columns.

3.3.3. C_{18} Preconditioning

The third column in the series is a small C_{18} column, which is used to collect cytokinins eluting from the DE52 column. Allow approx 0.4 mL BV C_{18} powder/g FW (*see* **Note 11**). Pack the C_{18} powder into a 5 or 10 mL syringe barrel. Precondition the column with 20 BV methanol, followed by 20 BV 10 m*M* ammonium acetate, pH 6.5.

3.3.4. Sample Application and Elution

After each column is packed and conditioned, connect the PVPP, DE52, and C_{18} columns in series with needles and bungs as shown in **Fig. 1** (*see* **Note 12**).

1. Add 20 mL (equivalent to 1 BV DE52) 10 m*M* ammonia acetate buffer, pH 6.5, to condition the entire complex.
2. Apply the pH-adjusted sample (2 mL) to the top column of the series through an attached 50-mL syringe barrel, which afterward serves as a solvent reservoir.
3. Elute the complex with 60 mL (equivalent to 3 BV DE52) 10 m*M* ammonium acetate buffer, pH 6.5, and then disconnect the three columns. Discard the PVPP column.

3.3.5. Nucleotide Recovery

1. Elute the DE52 column with 2.5 BV 1 *N* acetic acid (pH 2.5) in order to recover cytokinin nucleotides. Evaporate the nucleotide fraction (acid wash) to near dryness *in vacuo* (Savant Speed-Vac).
2. Redissolve the nucleotide fraction in 5 mL of the alkaline phosphatase reaction solution (*see* **Subheading 2.3.**) and subject to phosphatase treatment at 37°C overnight (approx 12 h) ***(27)***.
3. Load the reaction solution onto a second preconditioned C_{18} column (minimum 3 mL BV), and wash the column with 3 BV water. Elute with 3 BV 80% methanol to recover the dephosphorylated cytokinins. Evaporate the eluent to near dryness and redissolve in 50 µL 25% methanol prior to C_{18} HPLC.

3.3.6. Free base, Riboside, and Glucoside Recovery

Flush the C_{18} column with 80% methanol (3 BV) to elute free bases, ribosides, and glucosides. Dry the free base, riboside, and glucoside fraction and take up in 50 µL 50% acetonitrile prior to amine column HPLC (*see* **Subheading 3.4.1.**).

Table 1
Gradient Conditions for Normal Phase (NH_2) HPLC

Minute	Flow rate (mL/min)	%A[a]	%B
Initial	1.0	10	90
6	1.0	10	90
7	1.0	20	80
16	1.0	30	70
17	1.0	50	50
27	1.0	50	50
28	1.0	10	90

[a]Solvent A: Milli-Q water; solvent B: acetonitrile.

3.4. High Performance Liquid Chromatography

The plant sample is now purified sufficiently for the cytokinins to be separated using HPLC (Waters 600 Multisolvent controller, U6K injector port, Waters 490E Programmable multiwavelength UV detector set at 269 nm) (*see* **Note 13**).

3.4.1. Normal-Phase HPLC

Initially, a bulk separation of cytokinin free bases and ribosides from the glucosides is achieved using an amine column (HPLC Technology, Alphasil $5NH_2$ 5 µm, 250 × 4.6 mm) and a 25 min acetonitrile/water gradient ***(28)***.

1. Equilibrate the column with 10% H_2O: 90% acetonitrile.
2. *See* **Table 1** (*see* **Note 13**) for gradient conditions using a flow rate of 1 mL/min.
3. Wash the column with 50% H_2O: 50% acetonitrile and store in same.
4. The separation of the glucosides from the other cytokinins is verified using cytokinin standards; example retention times are shown in **Fig. 2**. Bulk collections are made from the HPLC. Fraction 1 (min 1–12) contains the free bases, ribosides, and iP9G, whereas fraction 2 (min 12–25) contains the remaining glucosides. Fraction 2 is evaporated to near dryness and treated with β-glucosidase reaction solution (*see* **Subheading 2.3.**) at 25°C for 22 h and run through

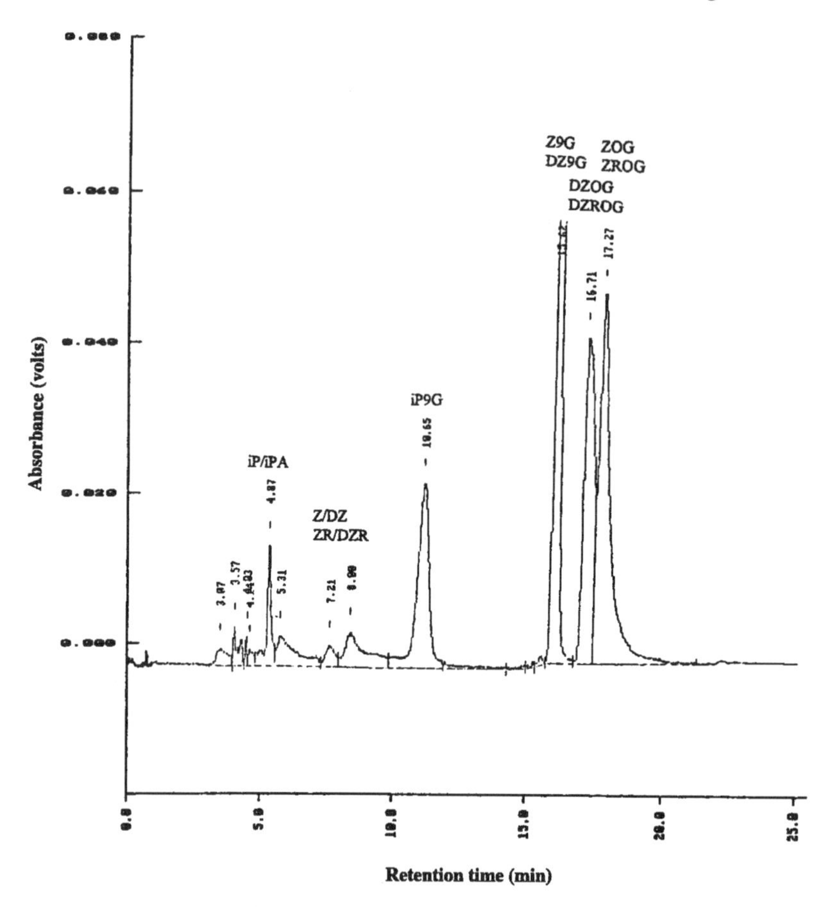

Fig. 2. Normal-phase HPLC separation of a mixture of cytokinin bases, ribosides, and glucosides. A clear cut at 12 min can be made to separate glucosides, excluding iP9G, from free bases and ribosides including iP9G. Column: Alphasil $5NH_2$, 5 µm, 250 × 4.6 mm. Detection: absorbance at 269 nm. *See* **Table 1** for details of the elution gradient.

another C_{18} column as described for nucleotide recovery (**Subheading 3.3.5.**). Both fractions are subsequently dried down ready for reverse-phase HPLC.

3.4.2. Reverse-Phase HPLC

The HPLC gradient is an adaptation of those described in MacDonald et al *(33)*, Jameson and Morris *(26)*, and Lewis et al.

Table 2
Gradient Conditions for Reverse-Phase (C_{18}) HPLC

Minute	Flow rate (mL/min)	%A[a]	%B	%C
Initial	1.0	75	25	0
1.0	1.0	75	25	0
5.0	1.0	72	28	0
9.5	1.0	61	39	0
10.0	0.8	61	39	0
15.0	0.8	60	40	0
15.5	1.0	60	40	0
20.0	1.0	59	41	0
21.0	1.0	75	0	25
22.0	1.0	75	0	25
26.5	1.0	71	0	29
31.5	1.0	70	0	30
37.5	1.0	0	100	0
47.5	1.0	0	100	0
48.5	1.0	75	25	0

[a]Solvent A: HPLC buffer; solvent B: methanol/HPLC buffer (80/20 v/v); solvent C: acetonitrile/HPLC buffer (80/20 v/v).

(28). Separation of individual cytokinins is achieved on an octadecyl silica C_{18} column (Beckman Ultrasphere 5 µm, 250 × 4.6 mm) with a triethylamine buffer (HPLC buffer: 40 m*M* acetic acid, pH 3.5, with triethylamine)/methanol/acetonitrile gradient.

1. Dissolve each of the fractions in 50 µL 25% methanol prior to injection on the HPLC.
2. Equilibrate the HPLC with HPLC buffer, pH 3.5, containing 20% methanol (*see* **Note 13**).
3. *See* **Table 2** for gradient conditions. We use a flow rate of 1 mL/min except for min 10–15, where it drops to 0.8 mL/min.
4. Collect fractions from the HPLC every 30 s into 1.5-mL polypropylene Eppendorf tubes, for the entire 40 min gradient during each sample run.
5. Sets of cytokinin standards (20 ng/cytokinin) are run through the column at the beginning and end of each session. This enables reten-

tion times of compounds exhibiting crossreactivity in the RIA to be compared with retention times of authentic standards (**Fig. 3**). Two blank gradients are run after the standards to prevent any crosscontamination between standards and samples (*see* **Note 14**).

6. Data is collected and processed on an IBM-compatible computer using Delta software (V. 4.06 Digital Solutions Ltd. Australia).

3.5. Radioimmunoassay

The streamlined protocol outlined below is based on more traditional RIA procedures ***(24,26)***.

1. Prepare the buffer/^{3}H-labeled antigen/antibody assay solution: 50 m*M* sodium phosphate buffer with 0.14 *M* NaCl, pH 7.2, 0.1% (w/v) gelatin (Difco), 0.01% ovalbumin (Sigma Grade V); ^{3}H-ZR-trialcohol or ^{3}H-iPA-trialcohol with radioactive strength being adjusted to 5000 CPMs per 450 μL reaction solution; and anticytokinin antibody providing approx 2500 CPMs binding (B_0)/450 μL reaction solution (*see* **Notes 15–18**).
2. Dissolve the dried fractions from C_{18} HPLC in 450 μL of the above buffer/antigen/antibody assay solution.
3. Leave samples for at least 4 h at room temperature with occasional stirring before adding 600 μL 90% saturated ammonium sulfate, pH 7.0, and mixing well. After 15 min centrifuge the tubes at approx 11,000*g* for 2 min. Aspirate the supernatants.
4. Release the antigen by addition of 50 μL methanol. Quantify the released radioactivity using liquid scintillation counting (Wallac) after addition of 1 mL scintillation cocktail.
5. In our work, fractions collected during C_{18} HPLC from 0 to 25.5 min are assayed with monoclonal antibody clone 16 (which crossreacts with hydroxylated cytokinins, including Z9G and DZ9G). ^{3}H-ZR-trialcohol is used as the competitor. Fractions eluting from 26 to 40 min are assayed with clone 12 (which crossreacts with nonhydroxylated cytokinins, including iP9G). ^{3}H-iPA-trialcohol is used as the competitor ***(28,29,34,35)***.
6. Standard curves should be included in each batch of RIAs. A series of dilutions of ZR or iPA between 0 and 100 pmol can be prepared by diluting a 1 mg/mL standard stock solution with 50% methanol into 1.5 mL Eppendorf tubes. Subsequently, dry the tubes using a Speed-Vac, then follow the steps in **Subheading 3.5.** starting from adding

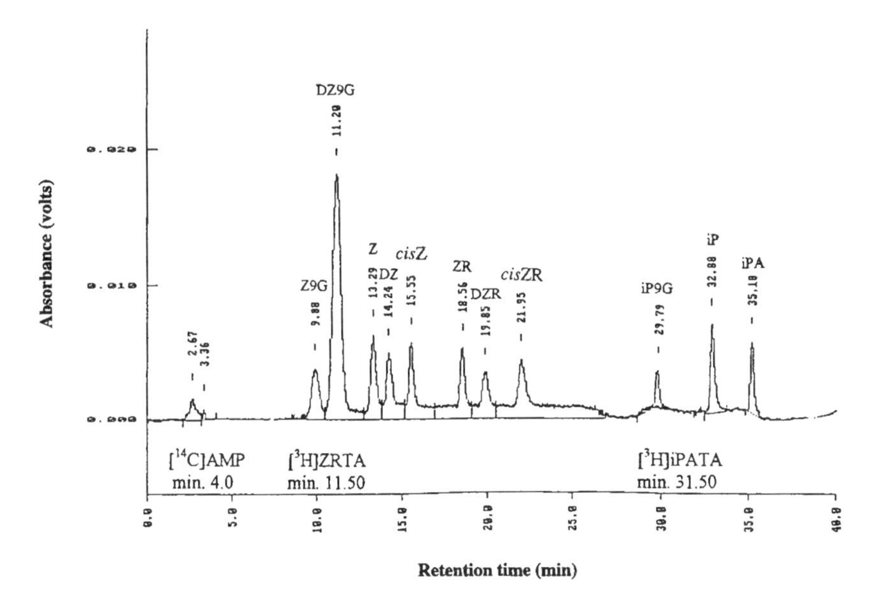

Fig. 3. Reverse-phase HPLC separation of mixed standards of naturally occurring cytokinins. The elution times of internal standards (^{14}C-AMP, ^{3}H-ZRTA, and ^{3}H-iPATA) are also shown. Column: Beckman Ultrasphere C_{18}, 5 μm, 250 × 4.6 mm. Detection: absorbance at 269 nm. *See* **Table 2** for details of elution gradient.

450 μL buffer/antigen/antibody assay solution. Triplicate tubes should be prepared for each concentration (*see* **Notes 19** and **20**).

4. Notes

1. Silanization of glassware: Rinse glassware with BDH dimethyldichlorosilane (2% in 1,1,1-trichloroethane) and remove any excess. After drying, wash the glassware with methanol and finally with water.
2. The protocol for the synthesis of ^{3}H-ZRTA or ^{3}H-iPATA can be found in MacDonald and Morris *(24)*. ^{14}C-AMP can be purchased from Amersham Life Sciences.
3. Protocols for the preparation of poly- and monoclonal antibodies can be found in MacDonald and Morris *(24)* and Banowetz *(29)*.
4. To obtain readily detectable amounts of cytokinins for RIA, we usually use 5–10 g FW of vegetative materials, but less of reproductive tissues. The detection limit of the RIA is approx 1 pmol ZR or iPA.

5. To prevent degradation caused by oxidation, antioxidants can be added to the sample prior to homogenization: e.g., 10 μL mercaptoethanol per milliter extraction solvent; other antioxidants may also be included *(28)*.
6. Tubes, such as 50 mL Falcon tubes, cannot be used as centrifuge tubes even at low speed because leaking occurs. However, they are suitable for use in a Savant Speed-Vac for evaporation of solvent.
7. For safety, we add the internal standards after homogenization. More correctly, they should be added during the initial extraction. ^{3}H-ZRTA and ^{3}H-iPATA elute separately from the endogenous cytokinins and can therefore be recovered for quantification of the ^{3}H without affecting the RIA (*see* **Fig. 3** for retention times). We sample 50 μL of HPLC eluent (tips used here should be soaked; *see* **Note 20**) from the appropriate fractions and mix with 1 mL scintillation cocktail.
8. Adjusting the pH of the sample: pH adjustment should be carried out when the volume is still large (as soon as the methanol is evaporated) and diluted alkaline (0.1 *N* NaOH) should be added very slowly and mixed quickly to prevent degradation caused by high temperature.
9. There is no need for a larger than 10 mL PVPP column here. A 5 mL BV of PVPP will generally be sufficient for removal of impurities and those solids sometimes produced while adjusting the pH of samples. A large PVPP column will reduce the elution rate from the column complex quite significantly.
10. Bulk DE52 can be preconditioned and may be stored at 4°C in the dark if 0.1% sodium azide is added. DE53 (twice the exchange capacity of DE52) or DEAE-Sephadex A25 (3.5 times the exchange capacity of DE52) *(36)* can be used if a large sample is to be extracted.
11. Prepacked C_{18} columns are commercially available, but more expensive. Filter-fitted polypropylene column reservoirs (10 mL) can be purchased. Place C_{18} powder in the reservoir up to a volume of 4 mL. If ordinary syringes are used extra care must be taken to prevent leaking, because C_{18} powder is very fine.
12. Multiple sets of the column complexes may be set up to process multiple samples simultaneously. We usually do 10 sets at one time. When connecting PVPP, DE52 and C_{18} columns, be careful not to disturb the column packings. Columns can be readily connected using methanol-washed bungs and syringe needles.
13. When HPLC is carried out, the injection loop should be washed with at least 100 μL methanol three times after each injection. Finally, wash the loop twice with solvent identical to the initial gradient before each sample or standard injection.

14. Blank gradients: Checking the blank gradients is important, because it will indicate if contamination has occurred after injection of the standards. In our laboratory, we usually collect the fractions from the second blank injection (50 µL of solvent identical to the solvent used to dissolve samples) onto the C_{18} column and then do a full RIA of the fractions.
15. RIA solution: Dissolve the gelatin in the buffer using a microwave; allow to cool before adding ovalbumin (this provides the base RIA buffer).
16. We usually handle between 400 and 800 Eppendorf tubes per assay. To calculate the volume of buffer needed for, say, 400 tubes, prepare 400 × 0.45 mL/tube = 180 mL plus some excess, e.g., to a total of 200 mL. Subsequently, add ^{3}H-ZRTA or ^{3}H-iPATA and adjust radioactivity to 5000 CPMs per 450 µL volume (as an additional check, always aliquot 450 µL out of the bulk solution and add 5 mL scintillation cocktail to check CPMs. It is not neccessary to have exactly 5000 CPMs/450 µL; a range between 4500 and 5000 CPMs is optimal).
17. When the antibodies are used for the first time, the antibody dilution needs to be determined at which the added amount (say 50 µL) of antibody binds 50% of the added ^{3}H-ZRTA or ^{3}H-iPATA in a 450 µL assay volume: use 350 µL of the base buffer, 50 µL ^{3}H-ZRTA or ^{3}H-iPATA containing 5000 CPM, and 50 µL antibody (use a series of dilutions of the original antibody stock). Once the appropriate dilution has been determined, the amount of the antibody stock required to be put in the bulk RIA solution may be calculated according to the formula:

$$X\ (\mu\text{L}) = \frac{(\text{RIA buffer volume [mL]}/0.45\ [\text{mL/tube}] \times 50\ \mu\text{L})}{(\text{Dilution at which 50 } \mu\text{L of antibody binds 50\% of the added radioactivity})} \tag{1}$$

    X is the amount of the antibody stock solution needed in the RIA solution.
18. The RIA solution can be stored at 4°C. Every time an RIA is to be carried out, the B_0 must be checked and a standard curve included with each particular batch of RIAs. B_0 is the radioactivity (CPM) bound to the antibody present in the 450 µL RIA buffer and which is precipitated by ammonium sulfate in the absence of other cytokinins. The ideal is to have approximately half the added radioactivity

bound to the antibody in the absence of other cytokinins. Before aliquoting the RIA solution into the sample tubes, it is very important to check B_0 by aliquoting 450 µL RIA buffer into three 1.5-mL Eppendorf tubes and then following the steps starting from adding the ammonium sulfate as described in **Subheading 3.5.**

19. Standard curve: Use B/B_0 as the Y axis and log (cytokinin standard pmol) as the X axis to draw the curve (B is the amount of radioactivity bound to the antibody in the presence of ZR or iPA standards).
20. All tips used to aliquot RIA solution, or in any other cases where accurately aliquoting small amounts of radioactivity or antibody is required, should be soaked with the solution for 15 min.

Acknowledgments

We acknowledge support from grants from the New Zealand Institute for Crop & Food Research (DHL) and from the New Zealand Forest Research Institute (HZ).

References

1. Weiler, E. W. (1979) Radioimmunoassay for the determination of free and conjugated abcisic acid. *Planta* **144,** 255–263.
2. Weiler, E. W. (1980) Radioimmunoassays for trans-zeatin and related cytokinins. *Planta* **149,** 155–162.
3. Hedden, P. (1993) Modern methods for the quantitative analysis of plant hormones. *Annu. Rev. Plant Physiol. Plant Mol. Bol.* **44,** 107–129.
4. Jameson, P. E. (1994) Cytokinin metabolism and compartmentation, in *Cytokinins: Chemistry, Activity and Function* (Mok, D. W. S. and Mok, M. C., eds.), CRC, Boca Raton, FL, pp. 113–128.
5. Tokota,T., Murofushi, N., and Takahashi, N. (1985) Extraction, purification, and identification, in *Hormonal Regulation of Development I: Molecular Aspects of Plant Hormones* (MacMillan, J., ed.), Springer-Verlag, Berlin, pp. 113–201.
6. Parker, C. W. and Letham, D. S. (1973) Regulators of cell division in plant tissues XVI. Metabolism of zeatin by radish cotyledons and hypocotyls. *Planta* **114,** 199–218.
7. Morris, J. W., Doumas, P., Morris, R. O., and Zaerr J. B. (1990) Cytokinins in vegetative and reproductive buds of *Pseudotsuga menziesii. Plant Physiol.* **93,** 67–71.

8. Dietrich, J. T., Kaminek, M., Blevins, D. G., Reinbott, T. M., and Morris, R. O. (1995) Changes in cytokinins and cytokinin oxidase activity in developing maize kernels and the effects of exogenous cytokinin on kernel development. *Plant Physiol. Biochem.* **33,** 327–336.
9. Bieleski, R. L. (1964) The problem of halting enzyme action when extracting plant tissues. *Anal. Biochem.* **9,** 431–442.
10. Horgan, R. and Scott, I. M. (1987) Cytokinins, in *Principles and Practices of Plant Hormone Analysis* (Rivier, L. and Crozier, A., eds.), Academic, New York, pp. 303–365.
11. Summons, R. E., Palni, L. M. S., and Letham, D. S. (1983) Determination of intact zeatin nucleotide by direct chemical ionisation mass spectrometry. *FEBS Lett.* **151,** 122–126.
12. Meilan, R., Horgan, R., Heald, J. K., LaMotte, C. E., and Schultz, R. C. (1993) Identification of cytokinins in red pine seedlings. *Plant Growth Regul.* **13,** 169–178.
13. von Schwartzenberg, K., Bonnet-Masimbert, M., and Doumas, P. (1994) Isolation of two cytokinin metabolites from the rhizosphere of Norway spruce seedlings (*Picea abies* L. Karst.). *Plant Growth Regul.* **15,** 117–124.
14. Jameson, P. E., Letham, D. S., Zhang, R., Parker, C. W., and Badenoch-Jones, J. (1987) Cytokinin translocation and metabolism in Lupin species. I Zeatin riboside introduced into the xylem at the base of *Lupinus angustifolius* stems. *Aust. J. Plant Physiol.* **14,** 695–718.
15. Wang, J., Letham, D. S., Taverner, E., Badenoch-Jones, J., and Hocart, C. H. (1995) A procedure for quantification of cytokinins as free bases involving scintillation proximity immunoassay. *Physiol. Plant.* **95,** 91–98.
16. Palmer, M. V., Horgan, R., and Wareing, P. F. (1981) Cytokinin metabolism in *Phaseolus vulgaris L.* : Identification of endogenous cytokinins and metabolism of [8-^{14}C] dihydrozeatin in stems of decapitated plants. *Planta* **153,** 297–302.
17. Muller, P. and Hilgenberg, W. (1986) Isomers of zeatin and zeatin riboside in clubroot tissue: evidence for trans-zeatin biosynthesis by *Plasmodiophora brassicae*. *Physiol. Plant.* **66,** 245–250.
18. Cappeiello, P. E. and King, G. J. (1990) Determination of zeatin and zeatin riboside in plant tissue by solid-phase extraction and ion-exchange chromatography. *J. Chromatog.* **504,** 197–201.
19. Biddington, N. L. and Thomas, T. H. (1976) Effect of pH on the elution of cytokinins from polyvinylpyrrolidone columns. *J. Chromatog.* **121,** 107–109.

20. Mousdale, D. M. A. and Knee, M. (1979) Poly-N-vinylpyrrolidone column chromatography of plant hormones with methanol as eluent. *J. Chromatog.* **177,** 398–400.
21. Parker, C. W. and Letham, D. S. (1974) Regulators of cell division in plant tissues XVII. Metabolism of zeatin in *Zea mays* seedlings. *Planta* **115,** 337–344.
22. Badenoch-Jones, J., Letham, D. S., Parker, C. W., and Rolfe, B. G. (1984) Quantitation of cytokinins in biological samples using antibodies against zeatin riboside. *Plant Physiol.* **75,** 1117–1125.
23. Guinn, G. and Brummett, D. L. (1990) Solid phase extraction of cytokinins from aqueous solutions with C_{18} cartridges and their use in a rapid purification procedure. *Plant Growth Regul.* **9,** 305–314.
24. MacDonald, E. M. S. and Morris, R. O. (1985) Isolation of cytokinins by immunoaffinity chromatography and analysis by high-performance liquid chromatography-radioimmunoassay. *Methods Enzymol.* **110,** 347–358.
25. Morris, R. O., Zaerr, J. B., and Chapman, R. W. (1976) Trace enrichment of cytokinins from Douglas-fir xylem exudate. *Planta* **131,** 271–274.
26. Jameson, P. E. and Morris, R. O. (1989) Zeatin-like cytokinins in yeast: Detection by immunological methods. *J. Plant Physiol.* **135,** 385–390.
27. Hocart, C. H., Letham, D. S., and Parker, C. W. (1990) Metabolism and translocation of exogenous zeatin riboside in germinating seeds and seedlings of *Zea mays*. *J. Exp. Bot.* **41,** 1517–1524.
28. Lewis, D. H., Burge, G. K., Schmierer, D. M., and Jameson, P. E. (1996) Cytokinins and fruit development in the kiwifruit (*Actinidia deliciosa*). I. Changes during fruit development. *Physiol. Plant.* **98,** 179–186.
29. Banowetz, G. M. (1994) Immunoanalysis of cytokinins, in *Cytokinins: Chemistry, Activity and Function* (Mok, D. W. S. and Mok, M. C., eds.), CRC, Boca Raton, FL, pp. 305–316.
30. Eason, J. R., Morris, R. O., and Jameson, P. E. (1996) The relationship between virulence and cytokinin production by *Rhodococcus fascians* (Tilford 1936) Goodfellow 1984. *Plant Pathol.* **45,** 323–331.
31. Pengelly, W. L. (1986) Validation of immunoassays, in *Plant Growth Substances 1985* (Bopp, M., ed.), Springer-Verlag, Berlin, pp. 35–43.
32. Prinsen, E., Redig, P., Van Dongen, W., Esmans, E. L., and Van Onckelen, H. A. (1995) Quantitative analysis of cytokinins by

electrospray tandem mass spectrometry. Rapid *Commun. Mass Spectrom.* **9,** 948–953.

33. MacDonald, E. M., Akiyoshi, D. E., and Morris, R. O. (1981) Combined high-performance liquid chromatography-radioimmunoassay for cytokinins. *J. Chromatog.* **214,** 101–109.
34. Trione, E. J., Krygier, B. B., Banowetz, G. M., and Kathrein, J. M. (1985) The development of monoclonal antibodies against the cytokinin zeatin riboside. *J. Plant Growth Regul.* **4,** 101–109.
35. Trione, E. J., Krygier, B. B., Kathrein, J. M., Banowetz, G. M., and Sayavedra-Soto, L. A. (1987) Monoclonal antibodies against the plant cytokinin isopentenyl adenosine. *Physiol. Plant.* **70,** 467–472.
36. Dawson, R. M. C., Elliott, D. C., Elliott, W. H., and Jones, K. M. (1986) Ion Exchange, gel filtration, and affinity chromatography media. In: *Data for Biochemical Research,* 3rd ed., Clarendon Press, Oxford, pp. 508–513.

8

Binding Studies

Michael A. Venis

1. Introduction

Each plant hormone elicits a set of characteristic physiological and biochemical responses. The primary stimulus initiating the sequence of events leading to such responses is the interaction of the hormone with its receptor protein. In general, receptors may be either membrane-associated or soluble (or solubilized) and this chapter deals with one binding technique suitable for each class. Some alternatives are briefly mentioned in **Subheading 4**, but for a comprehensive treatment of data acquisition and analysis, Hulme *(1)* is strongly recommended, together with the chapter by Klotz in **ref.** *2* on equilibrium dialysis.

In the simplest case, the reversible association of a hormone (H) with its receptor (R) is a bimolecular reaction:

$$H + R \underset{k_{-1}}{\overset{k_1}{\rightleftharpoons}} HR \tag{1}$$

where k_1 and k_{-1} are rate constants of association and dissociation, respectively. The equilibrium dissociation constant is given by:

$$K_d = \frac{k_{-1}}{k_1} = \frac{[H]\,[R]}{[HR]} \tag{2}$$

From: *Methods in Molecular Biology, vol. 141: Plant Hormone Protocols*
Edited by: G. A. Tucker and J. A. Roberts © Humana Press Inc., Totowa, NJ

K_d is equivalent to the hormone concentration at which half the binding sites are occupied (i.e., when $[R] = [HR]$) and, in cases where receptor occupancy is linearly coupled to response, it is also the concentration at which half-maximal biological response is achieved.

Since the total receptor concentration $[R_t] = [R] + [HR]$, then

$$K_d = \frac{[H]([R_t] - [HR])}{[HR]} \quad \text{or} \quad [HR] = \frac{[R_t][H]}{K_d + [H]} \tag{3}$$

Substituting standard binding terminology, this becomes:

$$B = \frac{nF}{K_d + F} \tag{4}$$

where $B = [HR]$ = concentration of bound hormone; $n = [R_t]$ = total receptor concentration; and $F = [H]$ = concentration of unbound (free) hormone.

Linear transformations of this equation or computerized curve-fitting procedures (e.g., **ref. *3***) are used to derive the desired binding constants K_d and n (***1,4***; **Note 1**). The calculation requires measurement, under equilibrium conditions, of B and F at a series of total hormone concentrations. The two methods described are based on those used for determining auxin binding to maize microsomal membranes ***(5)*** or solubilized preparations from these membranes ***(6)***. It is assumed that experimenters will have an appropriate particulate (membrane), cytoplasmic, or solubilized fraction as starting material and will tailor the precise conditions to the particular needs of their preparation and hormone. In the first method, radioactive hormone bound to particulate sites is separated from total hormone by centrifugation. The second method, equilibrium dialysis, involves the soluble protein fraction being placed on one side of a semipermeable membrane and dialyzed against labeled hormone. After equilibrium is reached, the value of F is obtained from the radioactivity in the protein-free solution, whereas the value of B comes from the difference in radioactivity between this and the protein-containing side, which represents $B + F$. In both the methods described the concentration of radioactive hormone is kept constant in the presence of increasing amounts of nonradioactive

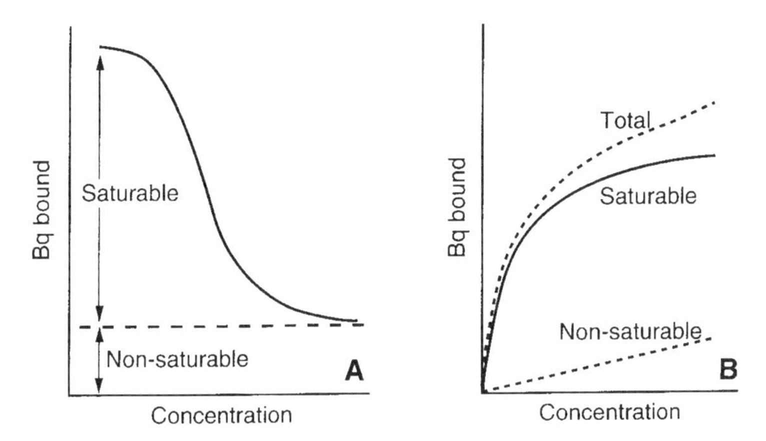

Fig. 1. Binding data obtained by **(A)** increasing the unlabeled hormone concentration at a constant, low concentration of radioactive hormone, or **(B)** increasing the radioactive hormone concentration in the presence (nonsaturable binding) or absence (total binding) of a fixed, high concentration of unlabeled hormone.

hormone (**Fig. 1A**). The alternative, if the radiolabel is not scarce or expensive, is to increase the concentration of radioactive hormone at constant specific activity in the presence or absence of a fixed, high concentration of unlabeled hormone (**Fig. 1B**). Radioactivity that is not displaced at a high unlabeled hormone concentration (**Fig. 1**) is termed "nonsaturable" binding and is subtracted from the total binding figure to yield the value for "saturable" or "specific" binding, i.e., *B*. Factors contributing to nonsaturable binding can be partitioning into lipid components, binding to low affinity, high abundance, nonspecific sites, or, in the case of membrane systems, occlusion in the membrane pellet.

2. Materials

1. Binding buffer: 0.25 *M* sucrose, 5 m*M* $MgCl_2$, and 10 m*M* trisodium citrate adjusted to pH 5.5 with acetic acid. Keeps for 7–10 d at 4°C (*see* **Note 2**).
2. The synthetic auxin naphthalene-1-acetic acid (NAA) is usually used as the ligand of choice, since in maize membranes the binding affinity is greater than that for the natural auxin indole-3-acetic acid

(IAA). Both NAA and IAA are obtainable from various suppliers, but may need recrystallizing from water. IAA should be protected from light both as a solid and in solution and stored cold. Stock solutions of NAA or IAA are prepared at 10 m*M* in methanol and stored at –20°C. A series of aqueous dilutions should be made at time of use to 100 times each required final concentration.

3. Radiolabeled auxins. NAA-^{14}C was originally used, but is no longer commercially available. NAA-^{3}H is prepared by catalytic tritiation of 4-bromo-NAA (synthesis in **ref.** *7*) and can be obtained by custom synthesis from Amersham International (Little Chalfont, UK) at a specific activity of about 1 TBq/mmol. Alternatively, IAA-^{3}H (Amersham, 0.5–1.1 TBq/mmol) can be used (*see* **Note 3**). Small amounts of working stock solutions are prepared by dilution in ethanol to 370 kBq/mL and stored at –20°C.
4. Water-compatible scintillation fluid, e.g., Ecoscint A (National Diagnostics, Atlanta, GA) or any other brand able to accommodate 1.5 mL of water/10 mL of scintillant.
5. For membrane fractions (centrifugation assay) we use polyallomer tubes of nominal 2 mL capacity from Kontron (Watford, Herts, UK) to fit four-place adaptors for their 8 × 50 mL rotor (i.e., a capacity of 32 tubes per rotor). *See* **Note 4** for alternatives.
6. For equilibrium dialysis we use equipment obtainable from Dianorm (Munich, Germany), Spectrum (Rancho Domínguez, CA), or NBS Biologicals (Hatfield, UK), with half-cells of 1 mL capacity, giving a capability of 20 assays/run. *See* **Fig. 2** and **Note 5** for other cell capacities and equipment. The Teflon half-cells are separated by a semipermeable membrane (Spectrapor 2, exclusion limit 12 kDa, 45 mm flat width, suppliers NBS Biologicals or Spectrum; other pore sizes available). *See* **Note 6**. The cell assemblies are held in a drive unit and rotated horizontally at up to 30 rpm.

3. Methods

3.1. Centrifugation Assay

1. All operations are carried out at 0–4°C. Using a Teflon-glass homogenizer, disperse the microsomal membrane pellet in binding buffer to give a homogenous suspension containing membranes from 0.5–0.7 g of original tissue fresh weight/mL of buffer. Measure out the requisite amount of buffer, use a spatula to dislodge pellets from

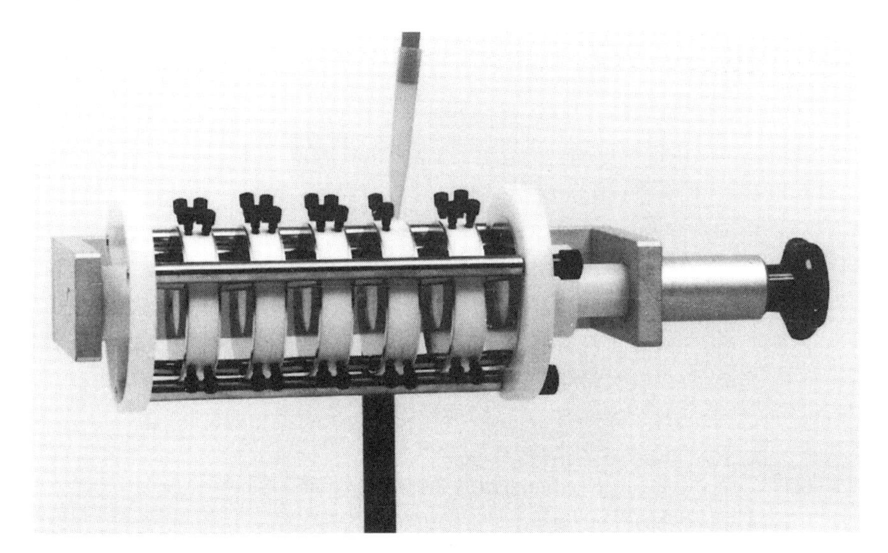

Fig. 2. Dialysis cell assembly (photo courtesy of Dianorm, Munich, Germany).

the centrifuge tubes in a few milliters of buffer, combine in the homogenizer, and rinse tubes with more buffer to complete the transfer. The Teflon-glass fit should be firm, but not overtight and full dispersion should need about 6–10 strokes, using enough buffer to fill the shaft (but not the cup) of the homogenizer. To avoid foaming, the Teflon plunger should remain within the liquid on each stroke. Decant the suspension to a conical flask and add the remainder of the buffer.

2. To the suspension add stock NAA-^{3}H (370 kBq/mL) at a ratio of 1–2 µL/mL. Count two or three 100-µL aliquots by liquid scintillation to determine total radioactivity input needed to calculate B at each total hormone concentration.
3. For each intended total hormone concentration, aliquot 3.2 mL radioactive suspension into a labeled glass 10-mL test tube using a variable 5 mL pipeter (*see* **Note 7**).
4. To each aliquot of radioactive suspension add 32 µL of unlabeled NAA from the dilution series (*see* **Subheading 2.2.**), to give an appropriate concentration range (for the NAA-maize system this is about 0.05–5 µ*M*, but may need to be varied for other systems; *see* **Note 3**). Briefly vortex-mix. Nonsaturable binding (*see* **Fig. 1A**) is determined from an aliquot with 32 µL of stock 10 m*M* NAA (final concentration 0.1 m*M*).

5. Dispense triplicate 1-mL aliquots from each 3.2-mL sample into labeled centrifuge tubes. Centrifuge at 13,000–20,000 rpm (20,000–48,000g_{max}) for 30–45 min (*see* **Notes 4** and **7**).
6. Aspirate supernatants completely, saving aliquots for scintillation counting if desired (*see* **Note 9**).
7. Carefully rinse tubes with two successive washes of about 1 mL cold water, using a Pasteur pipet to run the water down the side of the tube without disturbing the pellet.
8. After aspirating the second wash, add 450 μL water to each pellet, suspend carefully using a Pasteur pipet, and transfer to a scintillation vial, then rinse the tube with two further 450-μL washes to complete the transfer (*see* **Note 10**).
9. Add 10 mL scintillation fluid to each vial, shake vigorously, and count after equilibration for about 1 h.
10. Subtract the nonsaturable radioactivity value (at 0.1 m*M* NAA) from all other values to give saturable binding (*see* **Subheading 1.**).
11. From the resulting data, calculate *B* values at each *F*. *B* can be expressed as a concentration or as picomoles per milligram protein or per gram fresh weight. Derive K_d and *n* values graphically or by a curve-fitting program (*see* **Subheading 1.**).

3.2. Equilibrium Dialysis

1. Cut the edges off the roll of dialysis membrane, cut into squares, and soak in distilled water in a Petri dish. After a few minutes, each square can be separated into two membrane pieces using a finger and thumb at one corner, then peeling apart. Membranes should only be handled using disposable gloves and care should be taken to touch the membrane only at the corners.
2. To 22 mL cold binding buffer add 50 μL NAA-^{3}H (370 kBq/mL), then dispense ten 2.1-mL aliquots from this solution (*see* **Note 7**) into glass test tubes. Additionally, count 100-μL aliquots to determine label input (= *a* Bq/mL).
3. To each aliquot, add 42 μL unlabeled NAA from the dilution series (*see* **Subheading 2., item 2.**) (= *b* nmol/mL) and mix to give an appropriate concentration range (*see* **Subheading 3.1., step 4** and **Note 3**). To determine nonsaturable binding an aliquot with 42 μL of stock, 10 m*M* NAA is needed (giving a final concentration of

approx 0.1 m*M* after dialysis against an equal volume of the protein solution). Each 2.1 mL solution suffices for duplicate 1-mL assays (*see* **step 7**).

4. Dilute the soluble or solubilized binding protein preparation in cold binding buffer to give the equivalent of about 1 g of original tissue fresh weight/mL.
5. Assemble dialysis cells and membranes vertically in their carriers to give four sets of five cells (**Fig. 2**). For each cell, the base (or left) half is located on a steel spacer, a square of dialysis membrane is held at one corner and blotted by touching to tissue to remove excess water, then laid evenly across the cell cavity, and the unit is completed by laying the lid (or right) half cell on top. This is repeated five times and the assembled cells are secured in the carrier under slight tension. Store in a cold room until ready for filling.
6. Stopper the lower (emptying) ports of each half-cell and mount an assembled carrier horizontally in its stand (**Fig. 2**).
7. Into one of the upper ports of each right half-cell, pipet 1 mL radioactive hormone solution (from **step 3**), ensuring that the pipet tip seals into the port (**Fig. 2**). Stopper these ports.
8. Fill the corresponding left half-cells with 1 mL protein solution from **step 4** and stopper the ports.
9. Insert the remaining stoppers simultaneously in each half-cell pair to equalize pressure across the membrane (*see* **Note 11**).
10. Mount each completed assembly on the drive unit in the cold room and set to rotate at 20–30 rpm for 5 h or longer (*see* **Note 12**).
11. To unload, remount an assembly in its stand and rotate so that the unloading ports are above the horizontal. Remove the first unloading stopper, place the neck of a small glass test tube over the port, and rotate back to the vertical. On removing one of the upper stoppers, liquid will begin to flow out of the half-cell into the test tube and unloading is completed by forcing air through the port with a 1-mL pipeter. Repeat for the remaining half-cells.
12. Transfer 750 µL of each unloaded solution to a scintillation vial, add 10 mL of scintillation fluid, and count.
13. If Bq/mL in left cell = x and Bq/mL in right cell = y, and the mean difference between half-cells at 0.1 m*M* NAA = z Bq/mL, then $F = b\ y/a$ and $B = F\ (x - y - z)$.
14. *See* **Subheading 3.1., step 10** and **Note 13**.

4. Notes

1. **References *1*** and ***4*** should also be consulted for treatment of more complex binding situations.
2. The binding buffer recipe is that of Ray et al. ***(8)***. Other systems may well require different conditions, especially regarding pH, which should be tested initially in the range of 4.5–8.5 at unit intervals, using buffers of appropriate pK_a. Where possible, several buffers should be compared at each pH interval.
3. The specific activity of the hormone needs to be high enough to permit binding measurements at concentrations well below receptor saturation. Normally, 0.1–10 times the K_d is the useful range of measurements. Binding specificity can be evaluated by using auxin analogs in place of unlabeled NAA.
4. Ultracentrifuge rotors handling larger numbers of tubes are also available from Beckman, (Palo Alto, CA) (Type 25, 100 × 1 mL open tubes, up to $92{,}500g_{max}$; Type 50.4 Ti, 44 × 4 mL open tubes, up to $312{,}000g_{max}$; these can be used partly filled). The cheapest option, though far less satisfactory because of the lower centrifugal force obtainable (about $12{,}000g_{max}$), is to use a bench-top microcentrifuge and 1.5 mL Eppendorf tubes.
5. Half-cells of 0.2, 2, and 5 mL capacity are also obtainable. Other commercial units are available, e.g., from Hoefer (maximum half-cell volume = 0.5 mL). An inexpensive, though less convenient, system based on Eppendorf tubes has also been described ***(9)***. The simplest arrangement consists of knotted dialysis bags and is described in detail by Klotz ***(2)***.
6. The membrane width specified is suitable for Macro 1 cells (Dianorm). Other cells will required different width membranes.
7. It is good practice, in this and subsequent pipeting operations with radioactive solutions, to draw the solution up and down the pipet tip two to three times on first use, in order to saturate potential adsorption sites on the plastic. Otherwise, the first sample can be depleted in radioactivity relative to subsequent samples.
8. Equilibrium binding of NAA to maize microsomes is achieved very rapidly at 0–4°C and hence solutions can and should be centrifuged without delay. In other systems, binding may be time- and temperature-dependent. Initially, a comparison between 0° and 25°C incubation for 30 min will help to guide subsequent optimization.

9. This is only necessary in order to determine *F* in cases where ligand depletion is substantial. Normally, the amount bound is not more than 5% of *F*, in which case *F* can be simply equated to the total hormone concentration without materially affecting the derived K_d and *n* values.
10. It is important not to draw the suspension further up the pipet than necessary; otherwise pieces of pellet may become "stranded" and difficult to recover. The objective should be to transfer 90+% of the particulates in the first aliquot and the remainder in the second, leaving the third as a wash. With practice this is straightforward, but it is worth having a few trials first. Using 1 m*M* Tris base instead of water and leaving the first aliquot to sit for about 1 h tends to soften pellets and make transfer easier. Alternatively, instead of transferring the pellets, radioactivity can be extracted using methanol or ethanol *(8)*.
11. The filling operation for each five-cell assembly is sufficiently rapid that it can be carried out at the bench, using cold solutions.
12. If binding is temperature-dependent, dialysis can be carried out at room temperature, or the drive unit can be immersed in a thermostatted bath.
13. We have never encountered problems with binding of NAA-^{3}H to Teflon dialysis cells or to the membrane, but this needs to be checked for each ligand. If the protein carries a net charge, it may be necessary to counter Donnan effects by raising the ionic strength of the medium.

    At the end of an experiment the cells are disassembled and washed for subsequent use. Teflon cells are expensive and must be handled carefully to avoid damage to the rims that seal around the membrane. Lower the half-cells gently, cavity side up, into about 2 L of deionized water in a 5-L beaker, and leave for 15 min, swirling carefully once or twice. Decant the water and repeat the washing. Finally, rinse in 2 × 500 mL methanol, then leave to dry face-down on paper towels. The methanol washes are essential to ensure penetration into the narrow filling ports.
14. Notes on alternative binding techniques:
    a. For membrane or other particulate fractions, the centrifugation assay is the most reproducible, though filtration assays can be also be used (*see* Chapter 6 in **ref. *1***).
    b. Equilibrium dialysis is a rigorous technique, but since it relies on the difference between two fairly large numbers it is not the most sensititive for low levels of soluble binding. Alternatives include

pressure or centrifugal ultrafiltration, gel filtration, precipitation, and filter and adsorption assays; *see* **refs.** ***1*** and ***4*** for details and discussion of relative merits.

References

1. Hulme, E. C., ed. (1992) *Receptor-Ligand Interactions. A Practical Approach.* Oxford University Press, Oxford and New York.
2. Klotz, I. M. (1989) Ligand-protein binding affinities, in *Protein Function. A Practical Approach* (Creighton, T. E., ed.). Oxford University Press, Oxford, UK, pp. 25–54.
3. Munson, P. J. (1987) *A User's Guide to LIGAND.* Revised version 2.3.11. J. M. Coburn, Systex Inc., Beltsville, MD, USA.
4. Venis, M. (1985) *Hormone Binding Sites in Plants.* Longman, New York and London, pp. 24–40.
5. Batt, S., Wilkins, M. B., and Venis, M. A. (1976) Auxin binding to corn coleoptile membranes: kinetics and specificity. *Planta* **130,** 7–13.
6. Venis, M. A. (1977) Solibulisation and partial purification of auxin-binding sites of corn membranes. *Nature (Lond.)* **66,** 268,269.
7. Chandra, V. and Prasad, S. (1975) Condensation of 4-bromo-1-naphthylacetic acid with aldehydes. *J. Indian Chem. Soc.* **52,** 1223,1224.
8. Ray, P. M., Dohrmann, V., and Hertel, R. (1977) Characterization of naphthaleneacetic acid binding to receptor sites on cellular membranes of maize coleoptile tissue. *Plant Physiol.* **59,** 357–364.
9. Reinard, T. and Jacobsen, H.-J. (1989) An inexpensive small volume equilibrium dialysis system for protein-ligand binding assays. *Anal. Biochem.* **176,** 157–160.

9

Mutagenesis

Ottoline Leyser

1. Introduction

1.1. Why Make Mutants?

Mutants have been responsible for most of the recent achievements in plant hormone biology. Their importance cannot be overemphasized. The possession of lines that differ by only a single gene provides a powerful tool for following biosynthetic pathways, and for establishing the causal links between hormone and response that have been so hard to achieve by physiological and biochemical approaches. Moreover, the ability to clone a gene defined only by mutation has led to the isolation of many of the first genes known to be involved in hormone biosynthesis or signaling. The combination of mutants and cloned genes has opened the way for reverse genetic approaches that offer tremendous potential for the future. This chapter deals with some general considerations for the sucessful isolation of mutants and provides a mutagenesis protocol for *Arabidopsis*.

1.2. Genetic Study Systems

A number of plant species are particularly suitable for genetic studies (**Table 1**). The most appropriate for any particular applica-

From: *Methods in Molecular Biology, vol. 141: Plant Hormone Protocols*
Edited by: G. A. Tucker and J. A. Roberts © Humana Press Inc., Totowa, NJ

Table 1
Plants Suited to Genetic Studies

Plant	Life cycle	Growing space per plant	Classical genetics	Ease of transformation	Useful transposons	Molecular genetics
Moss	Vegetative	1 cm^2	±	+++	–	+
Pea	5 mo	75 cm^2	++	±	–	+
Rice	3 mo	10 cm^2	++	+	–	++
Antirrhinum	4 mo	75 cm^2	++	+	++++	+
Tobacco	4 mo	100 cm^2	+	++++	+	+
Petunia	4 mo	100 cm^2	++	++++	+	+
Maize	3 mo	900 cm^2	+++++	+	++++	+
Tomato	3 mo	600 cm^2	++++	+++	++	++
Arabidopsis	2 mo	1 cm^2	+++	++++	++	+++

tion will depend on a number of factors. Some are obvious biological considerations: The species used must reliably undergo the response in which you are interested. However, there are also several important pratical considerations: length of life cycle, ease of propagation (space and conditions required), genetic resources avaliable (maps, mutant collections, genome analysis), ease of transformation, and presence of active transposable elements.

Rice, moss, pea, petunia, *Antirrhinum*, and diploid tobacco species can provide excellent systems for genetic studies. Maize and tomato are even better suited to this approach. However, *Arabidopsis* is now established as the major genetic study system in plants and is likely to remain so because of its small size, short life cycle, and the ever expanding range of resources and support services available to *Arabidopsis* geneticists. Currently, unless you have a strong biological reason, it would be foolish to embark on a mutagenesis program using anything other than *Arabidopsis*. The protocol and associated notes given in **Subheadings 3** and **4** apply specifically to *Arabidopsis*, but the theoretical considerations are generally applicable and it would be straightforward to adapt the protocol for use with other plant species.

1.3. Screen Design

Once you have selected the appropriate species, the first step for a genetic approach is to design a screen for the mutant phenotype in which you are interested. Screens should be tested on a population of wild-type seed to assess space and labor requirements, as well as the extent of any environmentally induced variation in the character under examination.

The nature of the screen will have a major impact on the whole mutagenesis strategy. Obviously the number of possibilities is enormous. The easiest screens involve selection for vigorous seedling growth in conditions in which wild-type seedlings do not survive or grow only weakly, e.g., a hormone-resistance screen. Here, the desired mutants are easy to identify at a very early stage in the life cycle, minimizing both space and time requirements. However, it is important not to shy away from more complex, space- and time-consuming screens that can be a very shrewd investment.

Another important consideration in screen design is the selection of the genetic background for the mutagenesis. The use of already mutant backgrounds has proved a powerful approach (second-site suppresser or enhancer screens). For example, Abscisic Acid (ABA)-deficient mutants were identified because they can suppress the inability of Gibberellin (GA)-deficient mutants to germinate *(1)*. Similarly, transgenic backgrounds can be used. For example, a hormone-inducible promoter–reporter gene fusion can be used to identify mutants in hormone signaling. Even when wild-type backgrounds are used, there are usually several to choose from, some of which may be more suited than others to the screen in question.

1.4. The Basic Procedure

A flow chart for mutant generation is shown in **Fig. 1**. Standard plant protocols involve mutagenesis of dry seed. Plants germinating from these seeds are called the M1 generation. The progeny resulting from self-fertilization of the M1 are the M2 generation. Screening for mutants is usually carried out in the M2. This is because in the M1, mutations will be heterozygous and so only mutations dominant with respect to the wild-type allele will have a phenotypic effect. Furthermore, because the mutagenized embryo is a multi-celled structure, M1 plants are genetically mosaic. Only a small number of cells (the genetically effective cell number) from the embryo contribute to the germ line. It is therefore frequently the case that M1 phenotypes are not transmitted to the M2.

Plants selected from the M2 screen need some basic genetic analysis before investing time in detailed phenotypic characterization. The mutant phenotype should be retested in the M3 generation to eliminate false positives. If a large number of M2s are selected it is best to proceed to the M3 before making any crosses. If only a handful are selected, then crossing can start in the M2, saving time by a generation. M2 or M3 mutant plants should be back-crossed to the parental line, preferably as both a male and a female parent. The

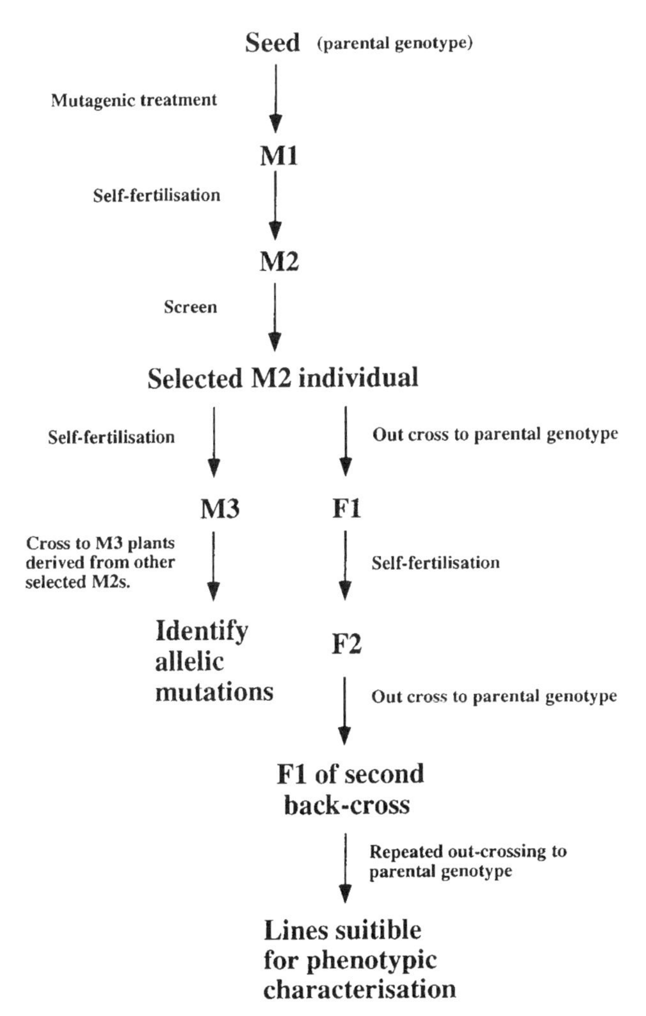

Fig. 1. Flow chart for mutant generation, selection, and basic genetic characterization.

resultant F1s will reveal the dominance relationship between the mutant and wild-type alleles. The segregation of mutant phenotypes in the F2s will indicate the number of loci involved, confirm the dominance relationships, and determine whether the mutation follows standard Mendelian inheritance patterns. The majority of mutant phenotypes selected are likely to result from single nuclear recessive loss-of-function mutations.

Mutant F2 individuals should be back-crossed again to the parental line in order to cross away any additional mutations not related to the mutation of interest. In theory, six rounds of back-crossing should be carried out to remove all except the most closely linked additional mutations. However, in practice three or four are usually sufficient to generate lines suitable for detailed phenotypic analysis. Crosses between different mutant isolates can be initiated to assign the mutants to complementation groups.

1.5. Saturation Mutagenesis

Usually the goal of a mutagenesis experiment is to identify all possible genes that can be mutated to give the desired phenotype. This approach is called saturation mutagenesis. When all possible genes have been hit, the screen is said to be saturated. Statistically, this is usually taken to be when the mean number of independent alleles recovered for each locus is five or greater. Obviously, there may still be additional loci that could be mutated to give the desired phenotype. Some genes are much more susceptible to mutation by one particular mutagen. It is therefore advisable to produce several M2 populations using different mutagens. Further screening may reveal new genes that, for example, only give the desired phenotype as a result of a rare dominant gain-of-function mutation. Here again the use of a range of mutagens will help to maximize the chances of identifying all possible genes.

Another important factor to consider is the pooling strategy for the M2. At one extreme, the seed from each M1 plant is collected separately and screened as an M2 family. This has the advantage that if allelic mutations are recovered from different M2 families, they are known to be independent events. Furthermore, in an M2 family screen, mutations that are homozygous lethal, and recessive with respect to wild-type can easily be recovered through their heterozygous sibs. Family screens are, however, very labor-intensive, and unless the mutations of interest are expected to be lethal, some pooling of the M2 seed is usual. In this case, pools of M1 plants are harvested together, generating a collection of M2 populations.

Mutations recovered from each pool are known to be independent and if the pools are derived from a relatively small number of M1 plants, there is still the possibility of reisolating homozygous lethal mutations through a limited M3 family screen.

The design of a mutagenesis experiment is usually a series of tradeoffs and compromises. The major limiting factors are space and time (both work time and absolute time). Although it would be better to be able to use many mutagens and to produce and screen enormous populations, this is usually not practicable. Instead careful consideration should be given to the experimental design in order to maximize the probability of finding all possible genes, while minimizing the total number of plants involved.

1.6. Choice of Mutagen

The probability of hitting a particular gene is obviously small, so many plants must be screened in order to identify individuals carrying a mutation in any one gene. In order to minimize the number of plants screened, the aim is to introduce as many mutations as possible into the genome, while maintaining high viability. The number of hits per genome is therefore an important parameter to consider in designing a mutagenesis. A second important feature is the spectrum of mutations introduced. Mutagens that induce point mutations give the widest mutational spectrum, since a point mutation can result in anything from complete loss of gene function to a subtle alteration in gene expression. This is particularly important for genes in which complete loss of function is lethal or, on the other extreme, is aphenotypic. Partial loss of function and dominant gain of function mutations allow the identification of such genes. Another advantage of point mutations is that the plant is likely to be able to tolerate more of them, compared to bigger deletions or rearrangement, without major viability problems in either the gametophytic or sporophytic generations.

A variety of mutagens is available, each of which has some advantages and some disadvantages. Using the criteria descibed above, chemical mutagens, such as ethyl methanesulfonate (EMS) or diepoxybutane (DEB), are undoubtedly the most efficient, and should usually be the first line of attack for a serious attempt at saturation mutagenesis.

Radiation and insertional mutagens can provide a useful adjunct to chemical methods, and are also particularly useful for so-called "targeted mutagensis," when mutation at a known locus or in a known sequence, rather than saturation mutagenesis, is the goal. Radiation (e.g., X-rays, γ-radiation, fast neutrons) has proven a useful mutagen for many systems. It is generally easy to use and effective. However, many radiation-induced mutations are relatively large-scale chromosomal rearrangements or deletions. This means that many genes may be affected by the same mutational event, transmissibility may be reduced, and the number of hits per genome is relatively low. Mutations can be induced by the insertion of a transposable element or a T-DNA. This has the advantage of providing a molecular tag with which to clone the mutationally identified gene. In the case of a transposon-induced mutation it may also allow the isolation of stable derivative alleles and form the basis for mosaic analysis experiments. The major drawbacks with insertional mutagensis are that both the number of insertions per genome and the mutational spectrum are relatively low.

1.6.1. Numbers

Many mathematical formulas allow the calculation of optimal numbers of M1 and M2 plants to produce and screen. Although these have their uses, they all depend on being able to estimate a variety of parameters, many of which are rather variable. It is therefore better to bear in mind the principles on which the formulas are built, while relying as much as possible on empirical data.

1.6.2. The M1

The optimal number of M1 plants depends on the number of mutations per genome induced by the mutagen, the size of the genome, the proportion of mutations that occur in genes vs intragenic regions, and the genetically effective cell number. Working back from empirical observations, it has been estimated that for an EMS mutagenesis of *Arabidopsis* an M1 of about 125,000 plants is likely to include all likely EMS-inducible mutations.

1.6.3. The M2 to M1 Ratio

The optimal ratio of M2 to M1 plants can depend on the frequency of mutation, the absolute size of the M2, and the genetically effective cell number. If the M2 is being screened in individual families and the genetically effective cell number is 2 (the best estimate for *Arabidopsis*) , then 20 plants per M1 parent must be screened for a 95% probability of identifying all the mutations in that family. In general, if the M2 pools are derived from relatively small M1 populations, then there is little to be gained by screening more than 20 M2s per M1 parent. In practice, unless your screen is very simple or your M1 is very small, this ratio will not be practical and more importantly, not necessary, to reach saturation of the screen using the criterion of five independent alleles per locus.

1.6.4. The M2

Clearly, up to a point, the more M2 plants you screen the better. However, even if your screen is particularly labor-intensive by careful choice of mutagen, and so forth, it is possible to achieve excellent results from a modestly sized M2. For example, in a typical EMS-mutagenized *Arabidopsis* M2, screening only 2000 plants will be sufficient to give a 95% probability of recovering one loss-of-function allele at any given locus.

2. Materials

The protocol described below is a highly serviceable minimal protocol. Some optional modifications are suggested in **Subheading 4.**

You will require:

1. Dry *Arabidopsis* seed of the appropriate genotype (*see* **Subheading 1.3.**). It is worth testing the viability of the seed batch before embarking on the mutagenesis.

2. EMS: EMS can be obtained from Sigma (St. Louis, MO) (listed as methanesulfonic acid, ethyl ester). It is a colorless volatile liquid and is highly mutagenic. EMS can be stored at –80°C, preventing evaporation. EMS solutions for mutagenesis should be made up in distilled water immediately before use. EMS should only be handled in a fumehood wearing suitable protective clothing. All equipment, glassware, and so forth that comes into contact with EMS should be soaked for 15 min in 100 m*M* sodium thiosulfate or 1 *M* sodium hydroxide, which destroy EMS, and then rinsed thoroughly in water. Solutions containing EMS can be decontaminated by adding solid sodium thiosulfate or sodium hydroxide.

   One option to give additional protection is to use an "AtmosBag" (available from Sigma). This is a sealable bag with built-in gloves (used in addition to appropriate protective gloves) into which all the equipment and solutions required for the mutagenesis are placed and sealed before the EMS bottle is opened. This is a particularly worthwhile precaution if the fumehood is likely to be used by others at any time during the mutagenesis. After the mutagenesis, the bag is opened, the decontaminated solutions are poured away and the decontaminated glassware and other items are left under running water to rinse.

3. Method

1. Place 20,000–250,000 dry seed in 50–100 mL of 30mM EMS (*see* **Note 1**).
2. Leave the seed at room temperature for 8–12 h, mixing occasionally.
3. Wash the seed twice for 15 min with 100 m*M* sodium thiosulfate to inactivate the EMS.
4. Wash the seed twice for 15 min with distilled water.
5. Sow the seed onto soil at a density of about 1 seed/cm^2 (*see* **Note 1**).
6. After about 2 wk, examine the trays and estimate the germination frequency. The germination frequency can be used to assess the effectiveness of the mutagenesis (*see* **Note 3**) and to give an estimate of M1 size.

4. Notes

1. Seed handling: The number of seeds used can be estimated by weight. One thousand Arabidopsis seeds weigh about 20 mg. During

the mutagenesis, several changes of solution are required. *Arabidopsis* seeds sink and it is possible to pour off most of the soaking solution with relatively little seed loss. The remainder can be removed with a transfer pipet. However, even this can be time-consuming and awkward, especially if working in an AtmosBag. To overcome this problem, the seeds can be kept in a syringe from which the end has been cut off and replaced with nylon mesh that can be melted onto the syringe. Solutions can be sucked up into the syringe and then expelled as required.

After the mutagenesis, the seed can be sowed immediately. An even distribution can be achieved by suspending the seed in a 0.1% agar solution and dispensing through a pipet. Alternatively, the seed can be dried on filter paper, scraped off, and mixed with dry sand. The sand–seed mixture can be sprinkled over the soil, giving an even distribution. Although a density of 1 seed/cm^2 is recommended, if the M1 plants are to be harvested individually a slightly lower density may be preferable. If pools of M1 plants are used, it is easiest to use a size of pot or seed tray that accommodates one M1 pool.

2. Variability: Considerable variation exists in the effectiveness of mutagenesis between experiments. This appears to relate to a range of factors, such as the batch of EMS, the ambient temperature, the quality of the seed, pH, and so on. Some of this variation can be reduced by the following optional extras *(2)*:
   a. Preimbibing the seed overnight in 10 m*M* potassium chloride.
   b. Making the EMS solution up in 100 m*M* sodium phosphate pH 5.0 and 5% dimethyl sulfoxide (DMSO).
   c. Minimizing temperature fluctuations.

   However, much of the variability seems to be in the EMS batch that is difficult to control. It may therefore be necessary to adjust either the EMS concentration or the time of exposure. One option is to use several EMS concentrations and then choose the best resulting M1.
3. Assessing the effectiveness of the mutagenesis: There are several ways in which to estimate the effectiveness of the mutagenesis. In the M1, indicators of a successful mutagenesis are a germination frequency of about 75% and the appearance of albino sectors on about 0.5% of M1 plants. Very high germination frequency or very infrequent sectoring indicates a poor mutagenesis. Conversely, too high a dosage may still give germination rates of 70% or more, but will render the M1 sterile. In a reasonable mutagenesis 10% of M1 plants should segregate embryo defects in the M2 (scoreable by examinaing

the siliques of M1 plants for one-fourth of the aborted seeds), and about 1% of M1 plants should segregate albinos in the M2. If your mutagenesis was effective by these criteria and you still do not get the mutants you had hoped for, then redesign your screen.

5. Commercial supplier of mutagenised seed: It is possible to buy mutagenized *Arabidopsis* seed from Lehle Seeds (P.O. Box 2366, Round Rock, TX78680-2366, USA; e-mail webmaster@arabidopsis.com; URL http://www.arabidopsis.com). They will also undertake custom mutageneses if you wish to use a nonstandard genetic background or mutagen. However, you will, of course, have more control and flexibility if you do it yourself.

References

1. Koornneef, M., Jorna, M. L., Brinkhorst-van der Swan, D. L. C., and Karssen, C. M. (1982) The isolation of abscisic acid deficient mutants by selection of induced revertents in non-germinating gibberellin sensitive lines of *Arabidopsis thaliana* (L) Heynh. *Theor. Appl. Genet.* **61,** 385–393.
2. Redei, G. (1969) *Arabidopsi thaliana* (L.) Heynh—A review of the genetics and biology. *Bibliogr. Genet.* **21,** 1–151.

Further Reading

More details and numerous references to good examples of successful mutagenesis programs can be found in relevant chapters of the following books:

Koncz, C., Chua, N.-H., and Schell, J., eds. (1992) *Methods in* Arabidopsis *Research.* World Scientific Publishing Co. Pte. Ltd.

Meyerowitz, E. M., and Somerville, C. R., eds. (1994) Arabidopsis. Cold Spring Harbor Laboratory Press, Cold Spring Harbor, NY.

Foster, G. D. and Twell, D., eds. (1996) *Plant Gene Isolation, Principles and Practice.* John Wiley and Sons, New York.

10

The Identification of Ethene Biosynthetic Genes by Gene Silencing

Antisense Transgenes, Agrobacterium*-Mediated Transformation, and the Tomato ACC Oxidase cDNA*

Grantley W. Lycett

1. Introduction

1.1. Antisense and Sense Gene Silencing

Artificial oligodeoxynucleotides *(1,2)*, and later RNA *(3)* complementary to a particular gene, were shown to inhibit the expression of that gene in vertebrate cells. Transformation of plants with antisense transgenes was first used to block the expression of well-characterized plant genes in 1988 *(4–6)* and later it was found that sense transgenes would often similarly block endogenous gene expression *(7–9)*. More recently, it has been shown that a single chimeric gene consisting of coding regions from two endogenous genes can silence the expression of both of the endogenous genes *(10)*. This can be very useful if one of the two sequences is from a gene that gives rise to a color or other visual phenotype.

In many cases, genes, or more often cDNAs, may be cloned and sequenced, and little may be learned about the function of the gene

From: *Methods in Molecular Biology, vol. 141: Plant Hormone Protocols*
Edited by: G. A. Tucker and J. A. Roberts © Humana Press Inc., Totowa, NJ

product by computer homology searches and other sequence-based predictive methods. In such cases, antisense transformation may be used to block the expression of the gene to create a null-mutant phenotype. The nature of the phenotype may then indicate something of the nature of the biochemical process in which the gene product is involved.

1.2. ACC Oxidase

The first unknown cDNA to be characterized by antisense transformation was pTOM13. This was a cDNA isolated by virtue of its induction during the ripening of tomato fruit ***(11)*** and subsequently shown to be induced by wounding ***(12,13)***. There was no known function assigned to this gene; however, because both of these processes involve ethene production there was circumstantial evidence that this cDNA might be either induced by ethene or involved in ethene biosynthesis. Elucidation of the actual role came from antisense transformation of tomato to produce plants with a reduced ability to synthesize ethene ***(14)***. These plants also had a reduced ability to convert the immediate precursor of ethene, 1-aminocyclopropane-1-carboxylic acid (ACC) into ethene when supplied exogenously, indicating that pTOM13 encoded ACC oxidase (ACO), sometimes called ethylene-forming enzyme (EFE).

Historic attempts to purify the ACC oxidase enzyme had always been largely unsuccessful. However, identification of the ACO cDNA allowed sequence comparisons to be made with other enzymes and showed that ACO had amino-acid identity with a flavone-3-hydroxylase of *Antirrhinum majus* (Prescott and Martin in **ref.** ***14***). The knowledge that this hydroxylase requires ascorbate and iron allowed Ververidis and John ***(15)*** to purify and assay ACO using the same factors.

Confirmation of the identity of pTOM13 came from experiments where the pTOM13 coding sequence was expressed in yeast, allowing yeast to convert ACC to ethene ***(16)***. This conversion was stereoisomer-specific, and was stimulated by iron and ascorbate.

1.3. ACO Antisense Plants

The production of ACO antisense plants has had a number of uses. First, the fruit of ACO antisense plants ripen normally but have a much increased shelf life because of inhibition of overripening and reduced mechanical and fungal damage. The life of cut flowers is similarly extended. Second, the plants have been used to study phenomena thought to be induced by ethene, such as senescence ***(17)***, epinasty ***(18)***, and pathogen attack ***(19)***.

2. Materials

2.1. Gene Construction

1. Plasmid vectors: As source of expression cassette, pDH51 ***(20)***.
2. *Agrobacterium* binary transformation vector, pBIN19 ***(21)***.

2.2. Agrobacterium-*Mediated Transformation*

1. APM-strep medium: 0.5% (w/v) yeast extract, 0.05% (w/v) casamino acids, 0.8% (w/v) mannitol, 0.2% (w/v) $(NH_4)_2SO_4$, 0.5% (w/v) NaCl, pH 6.6, containing 500 µg/mL streptomycin.
2. APMSS medium: As APM-strep but additionally substituted with kanamycin at 50 µg/mL (or tetracycline at 12.5 µg/mL, as appropriate for the plasmid used).
3. MS medium: 4.4 g/L MS salts, 30 g/L sucrose.
4. MSR3 agar: 4.4 g/L MS salts, 30g/L sucrose, 1 mg/L thiamine HCl, 0.5 mg/L nicotinic acid, 0.5 mg/L pyridoxine HCl, adjusted to pH 5.9 with KOH and autoclaved with 8 g/L agar. For convenience, the thiamine, nicotinic acid, and pyridoxine may be prepared as a 1000 × concentrated solution (R3 vitamins).
5. MS24D medium: As MSR3 without the agar but with 200 mg/L KH_2PO_4, 100 µg/L kinetin, and 200 µg/L 2,4-D and adjusted to pH 5.7.
6. 3C5ZR agar: As MSR3 agar with 1.75 mg/L zeatin riboside and 0.87 mg/L indoleacetyl-DL-aspartic acid added after autoclaving.
7. 3C5ZCK agar: As 3C5ZR agar with 400 mg/L augmentin and 75 mg/L kanamycin sulfate.

2.3. Analysis of Plants

1. 10 × PCR buffer: 0.1 *M* Tris-HCl pH 8.0, 0.5 *M* KCl, 20 m*M* MgCl, 0.01% (w/v) gelatin, 0.5% (v/v) Nonidet P-40, 0.05% (v/v) Tween 20.
2. dNTP mix: 2.5 m*M* each of dATP, dCTP, dGTP and TTP, filter-sterilized. It is convenient to make up as separate 100 m*M* stocks.

2.4. Ethene Assay and Simple In Vivo ACC Oxidase Assay

1. Subaseal™ caps (W.H. Freeman and Co., Barnsley, UK)
2. Ethene standard (100ppm)
3. ACC solution: 1 m*M* ACC, 30 m*M* Na ascorbate, 0.1 m*M* $FeSO_4$, 0.8 *M* mannitol, 10% glycerol, 0.1 *M* Tris-HCl, pH 7.2.

3. Methods

3.1. Gene Construction

The main problem with gene construction is the correct choice of vector and coding sequence. We have generally used pBIN19 as the vector in *Agrobacterium tumefaciens* strain LBA4404 and have had good results. pDH51 is the usual source of promoter and polyadenylation site sequences. In order to produce a fragment with the correct terminal restriction enzyme sites to allow it to be inserted in the correct orientation, several intermediate cloning steps may need to be performed. Fortunately this is not as great a problem as one would assume. It is not necessary, or even desirable, to use a whole gene or even a whole cDNA for gene silencing. Fragments of cDNA between 200 and 1000 nucleotides generally work well. There is no generally accepted way of predicting which fragment will give the most efficient silencing. However, very small fragments are unlikely to work well and very large ones will add to the difficulties of subcloning and transformation. Complete sequence identity is not necessary and, in the case of multigene families, it is probable that more than one endogenous gene will be silenced by the transgene, as in the case of the highly conserved ACC oxidase genes. It is not always necessary to use a gene sequence from the same species if the gene in question is highly conserved between species.

It is difficult to choose whether to use antisense or sense transgenes. We have found that antisense and sense silencing work similarly in efficiency and perhaps mechanism *(22)*. One advantage of using sense transgenes consisting of the full coding sequence is that silencing and overexpression may occur in different plants within the same population of transformants. Overexpressers may then be useful controls.

Some transgenic plants are mosaics of silenced and nonsilenced cells. Downregulation of the phytoene synthase gene causes a dramatic change in color of tomato fruit *(23)*. Because chimeric transgenes always seem to silence both of the endogenous genes from which they were made, by fusing the *Psy* coding sequence to part of the polygalacturonase cDNA it was possible to identify those plants that were mosaics and those tissues where the *Pgu* gene was silenced by virtue of the colour change caused by silencing of the *Psy* gene *(22)*. It may be possible to use other such systems to identify such mosaic tissues (*see* **Note 1**).

3.2. Agrobacterium*-Mediated Transformation*

This is the procedure used in our laboratories for the routine transformation of tomato (*Lycopersicon esculentum* Mill. cv Ailsa Craig) with pBIN19 in *Agrobacterium tumefaciens* strain LBA4404. In contrast to certain other procedures, such as promoter analysis, antisense or sense gene silencing does not require large numbers of transformants. This method has worked well in tomato but because transformation procedures are notoriously species-specific, a modification of the procedure or a different procedure altogether may be needed for different types of plants.

3.2.1. Transformation of Agrobacterium

1. To prepare competent cells, strain LBA4404 (which contains a helper Ti plasmid) should be streaked onto an APM-strep agar plate and grown at 29°C for several days in the dark (*see* **Note 2**). Subculture the bacteria into APM-strep liquid medium at 29°C overnight in the dark. The following day, dilute 2 mL of the culture into 50 mL of

fresh APM-strep medium in a 100-mL flask and grow with vigorous shaking to A_{600} of 0.5–1.0.

2. Chill the culture on ice, centrifuge, and resuspend the cell pellet in 1 mL of ice-cold calcium chloride (20 m*M*). Dispense 100-μL aliquots into chilled 1.5-mL microcentrifuge tubes and use for transformation.
3. To an aliquot of competent cells, add 1 μg plasmid DNA, followed by 1 mL of APM medium and incubate at 29°C with gentle shaking for 2–4 h in the dark.
4. Centrifuge the cells (in a microcentrifuge for only 30 s) and resuspend in 100 μL of fresh APM medium. Spread on APMSS plates and incubate the plates at 29°C for several days in the dark until colonies appear.

3.2.2. Transformation of Tomato

1. Inoculate a culture of *Agrobacterium* into APM medium with antibiotics 5 or 6 d before it is to be used and grow as before. This should be diluted 10^{-2} in fresh APM 3 d before use. Dilute 0.2 mL into 10 mL fresh APM again 16 h before use.
2. Surface-sterilize seeds by soaking for 10 min in 50% (v/v) ethanol and 20 min in saturated trisodium *ortho*phosphate (*see* **Note 3**), and washing briefly in sterile distilled water. They should then be soaked for 10 min in 50% (w/v) sodium hypochlorite, rinsed three times in sterile distilled water, and sown (approx 150 per pot) on sterile MSR3 agar. Incubate at 26°C with 16 h light and 8 h dark.
3. Cut cotyledonary explants (*see* **Note 4**) from 10–12-d-old seedlings and incubate in MS24D medium for 20 min.
4. Place on sterile filter paper laid on 3C5ZR plates (*see* **Note 4**). The plates should be sealed with film and incubated in low light (e.g., under a layer of muslin) for 24 h.
5. Centrifuge 25 mL of overnight *Agrobacterium* culture (*see* **Note 5**), wash the pellet in APM medium without antibiotics, and resuspend it in MS medium. Prepared cotyledonary explants are added to the culture and shaken very gently for 15–20 min.
6. Remove the explants and blot dry on sterile filter paper, then replace the filter onto the 3C5ZR plates and incubate as in **step 4** for 2 d. Transfer the explants to 3C5ZCK plates. Keep in low light for the first week and then transfer to normal light. Continue to transfer the explants to fresh plates every 3 wk until calli and/or shoots appear.

7. At this point, remove dead tissue, divide the callus into approx 1.5 cm^2 sections, and put into 3C5ZCK tubs, which are again subcultured every 3 wk. Shoots should be excised and cultured in MSR3CK tubs until roots form, then grown in compost.

3.3. Analysis of Plants

As with any transgene, those used for gene silencing will not always be expressed efficiently. It is generally assumed that this is associated with chromosomal location. In addition, gene silencing is an epigenetic effect and the phenotype is not always stable, especially through meiosis. In the case of the ACC oxidase antisense plants, it was possible to test for efficient expression of the transgene in the leaves by northern blot because the transgene promoter was constitutive and the promoters of the endogenous genes were not. In the ripe fruit, RNA from the transgene could not be detected because of the mutual nature of gene cosuppression ***(14)***. Expression of a transgene does not always silence the endogenous genes, so the use of a chimeric transgene containing a visual reporter may be advantageous. This may also be useful if the object is to silence one member of a multigene family. We have found that it is sometimes possible to silence a gene and obtain a phenotype but still to see bands in northerns because of the continued expression of other homologous genes. Ultimately, it may be necessary to show that the phenotype obtained cosegregates with the presence of the transgene. The possibility of insertional inactivation may be excluded by demonstrating that the same phenotype occurs in several independent transformants.

3.3.1. PCR

1. Add 50 µL of mineral oil to the open lid of a 0.5 mL microcentrifuge tube.
2. Mix 5 µL 10 × PCR buffer, 4 µL 2.5 m*M* dNTP mixture, 2.5 µL 20 µ*M* primer 1, 2.5 µL 20 µ*M* primer 2, 30.8 µL sterile H_2O, and 0.2 µL Taq DNA polymerase in the tube (*see* **Note 6**).
3. Add 5 µl of template DNA (at about 100–200 µg/mL).

4. Amplify for about 35 cycles in a thermal cycler. The exact conditions for this will depend on the cycler.

3.4. Ethene Measurements and ACC Oxidase Assay

Effective assays for purified ACC oxidase are now available. However, we have found that measurements of ethene production or a simple assay of ACC oxidase activity in vitro are normally quite adequate.

3.4.1. Measurement of Ethene Production

1. Cut tissue sections and place in 5-mL glass bottles sealed with self-sealing rubber caps (*see* **Note 7**).
2. Remove 1-mL samples of air from the headspace using a gas-tight syringe at regular intervals of 30–90 min (*see* **Note 8**) and aerate and reseal the bottle after each sampling.
3. Samples should be measured on a gas chromatograph. We use a Pye Unicam apparatus with an alumina column and a flame ionization detector. An ethene source of known concentration should be used as a standard.

3.4.2. ACC Oxidase Assay In Vitro

1. Incubate tissue samples, prepared as in **Subheading 3.4.1.**, in ACC solution *(15)* for 15 min and discard the solution.
2. Seal the samples into glass bottles and assay as in **Subheading 3.4.1.**

4. Notes

1. Oeller and Gutterson *(24)* have described a somewhat similar system that is more generally applicable but is less immediate in that it requires a tissue-staining step. In their system, a sense *uidA* (GUS) reporter is fused to a sense transgene. In the absence of gene silencing, readily detectable glucuronidase will be formed. If cosuppression occurs, not only will the endogenous genes be suppressed, but the transgene message will also be degraded and no glucuronidase enzyme will be made.
2. Incubation of *Agrobacterium* above 29°C or in the light is not recommended because loss of the plasmid is more likely.
3. This is an antiviral treatment.

4. This is best achieved by making transverse cuts across the two ends of the cotyledon. Cut ends of the explants should be in contact with the filter paper or the agar, and the natural curve of the cotyledons will assist this.
5. It is important to check the density of the culture. The A_{600} should be between 0.5 and 0.8.
6 For greater accuracy of pipeting, if several reactions are to be performed, make up a larger quantity (slightly more than needed) and dispense 45-µL aliquots into the microcentrifuge tubes from this.
7. Various types of tissue section may be used. Tomato leaf disks cut with a sharp cork borer (5 per bottle) and 2-cm petiole sections (2 per bottle) have both proved suitable.
8 This method will cause wounding of the plants and induction of wound ethene production. If desired, this may be circumvented by taking samples quickly, before wound induction has taken place. The first 30 min has proven a suitable period for our studies on tomato petioles *(18)*. To avoid wounding, whole fruit may be placed in larger (380-mL) glass jars and assayed in the same way.

Acknowledgments

I am very grateful to Chungui Lu and Craigh Jones for discussing their methods and making helpful criticisms of the manuscript.

References

1. Zamecnik, P. and Stephenson, M. (1978) Inhibition of Rous sarcoma virus replication and cell transformation by a specific oligodeoxynucleotide. *Proc. Natl. Acad. Sci. USA* **75,** 280–284.
2. Stephenson, M. and Zamecnik, P. (1978) Inhibition of Rous sarcoma viral RNA translation by a specific oligodeoxyribonucleotide. *Proc. Natl. Acad. Sci. USA* **75,** 285–288.
3. Izant, J. G. and Weintraub, H. (1984) Inhibition of thymidine kinase gene expression by anti-sense RNA—a molecular approach to genetic analysis. *Cell* **36,** 1007–1015.
4. van der Krol, A. R., Lenting, P. E., Veenstra, J., van der Meer, I. M., Koes, R. E., Gerats, A. G. M., Mol, J. N. M., and Stuitje, A. R. (1988) An anti-sense chalcone synthase gene in transgenic plants inhibits flower pigmentation. *Nature* **333,** 866–869.

5. Smith, C. J. S., Watson, C. F., Ray, J. Bird, C. R., Morris, P. C., Schuch, W., and Grierson, D. (1988) Antisense RNA inhibition of polygalacturonase gene expression in transgenic tomatoes. *Nature* **334,** 724–726.
6. Sheehy, R. E., Kramer, M., and Hiatt, W. R. (1988) Reduction of polygalacturonase activity in tomato fruit by antisense RNA. *Proc. Natl. Acad. Sci. USA* **85,** 8805–8809.
7. Napoli, C., Lemieux, C., and Jorgensen, R. (1990) Introduction of a chimeric chalcone synthase gene into petunia results in reversible co-suppression of homologous genes *in trans*. *Plant Cell* **2,** 279–289.
8. van der Krol, A. R., Mur, L. A., Beld, M., Mol, J. N. M., and Stuitje, A. R. (1990) Flavonoid genes in petunia: addition of a limited number of gene copies may lead to a suppression of gene expression. *Plant Cell* **2,** 291–299.
9. Smith, C. J. S., Watson, C. F., Bird, C. R., Ray, J., Schuch, W., and Grierson, D. (1990) Expression of a truncated tomato polygalacturonase gene inhibits expression of the endogenous gene in transgenic plants. *Mol. Gen. Genet.* **224,** 477–481.
10. Seymour, G. B., Fray, R. G., Hill, P., and Tucker, G. A. (1993) Down-regulation of two non-homologous endogenous tomato genes with a single chimaeric sense gene construct. *Plant Mol. Biol.* **23,** 1–9.
11. Slater, A., Maunders, M. J., Edwards, K., Schuch, W., and Grierson, D. (1985) Isolation and characterisation of cDNA clones for tomato polygalacturonase and other ripening-related proteins. *Plant Mol. Biol.* **5,** 137–147.
12. Smith, C. J. S., Slater, A., and Grierson, D. (1986) Rapid appearance of an mRNA correlated with ethylene synthesis encoding a protein of molecular weight 35 000. *Planta* **168,** 94–100.
13. Holdsworth, M. J., Bird, C. J., Ray, J., Schuch, W., and Grierson, D. (1987) Structure and expression of an ethylene-related mRNA from tomato. *Nucleic Acids Res.* **15,** 731–739.
14. Hamilton, A. J., Lycett, G. W., and Grierson, D. (1990) Antisense gene that inhibits synthesis of the hormone ethylene in transgenic plants. *Nature* **346,** 284–287.
15. Ververidis, P. and John, P. (1991) Complete recovery *in vitro* of ethylene-forming enzyme activity. *Phytochemistry* **30,** 725–727.
16. Hamilton, A. H., Bouzayen, M., and Grierson, D. (1991) Identification of a tomato gene for the ethylene-forming enzyme by expression in yeast. *Proc. Natl. Acad. Sci. USA* **88,** 7434–7437.

17. John, I., Drake, R., Farrell, A., Cooper, W., Lee, P., Horton, P., and Grierson, D. (1995) Delayed leaf senescence in ethylene-deficient ACC-oxidase antisense tomato plants - molecular and physiological analysis. *Plant J.* **7,** 483–490.
18. English, P. J., Lycett, G. W., Roberts, J. A., and Jackson, M. B. (1995) Increased 1–aminocyclopropane-1–carboxylic acid oxidase activity in shoots of flooded tomato plants raises ethylene production to physiologically active levels. *Plant Physiol.* **109,** 1435–1440.
19. Cooper, W., Bouzayen, M., Hamilton, A., Barry, C., Rossall, S., and Grierson, D. (1998) Use of transgenic plants to study the role of ethylene and polygalacturonase during infection of tomato fruit by *Colletotrichum gloeosporioides. Plant Pathol.* **47,** 308–316.
20. Pietrzk, M., Shillito, R. D., Hohn, T., and Potrykus, I. (1986) Expression in plants of two antibiotic resistance genes after protoplast transformation with a new plant expression vector. *Nucleic Acids Res.* **14,** 5857–5868.
21. Bevan, M. W. (1984) Binary *Agrobacterium* vectors for plant transformation. *Nucleic Acids Res.* **12,** 8711–8721.
22. Jones, C. G., Scothern, G. P., Lycett, G. W., and Tucker, G. A. (1998) Co-ordinated gene silencing by a single chimeric DNA construct. *Planta* **204,** 499–505.
23. Bird, C. R., Ray, J. A., Fletcher, J. D., Boniwell, J. M., Bird, A. S., Teulieres, C., Blain, I., Bramley, P. M., and Schuch, W. (1991) Using antisense RNA to study gene function: inhibition of carotenoid biosynthesis in transgenic tomatoes. *Biotechnology* **9,** 635–639.
24. Oeller, P. W. and Gutterson, N. (1997) A generic assay for gene silencing. 5th International Congress of Plant Molecular Biology, Singapore, September 1997, Abstract No 975.

11

Extraction, Separation, and Analysis of Plant Phosphoinositides and Complex Glycolipids

Bjørn K. Drøbak, Nicholas J. Brewin,
and Luis E. Hernandez

1. Introduction

1.1. Lipid Signaling

Progress in the understanding of how cells perceive signals and how the perception of specific signals leads to cellular responses has been dramatic over the last decade. One of the many new aspects of cell signaling that has emerged is that several types of lipids play key roles in the transfer of information across biological membranes. In mammalian cells it has long been suspected that a certain group of inositol-containing lipids, known collectively as the phosphoinositides, is linked to transmembrane signaling *(1)* but it is only in the last 10–15 yr that sufficient evidence has been provided to establish that the phosphoinositide signaling system *(2,3)*, and other lipid signaling cascades *(4)*, are essential components of a very wide range of signaling casettes in eukaryotes. The bewildering complexity of lipid signaling pathways can appear daunting to newcomers to the field, and may sometimes result in researchers electing to avoid entering into experimentations in this field. It is not our intention to present an exhaustive catalog of methods for the study

From: *Methods in Molecular Biology, vol. 141: Plant Hormone Protocols*
Edited by: G. A. Tucker and J. A. Roberts © Humana Press Inc., Totowa, NJ

of lipid signaling components and how these may be affected by plant hormones/growth factors; rather we will present a few methods that have been tried and tested in our laboratories, are fairly straightforward to set up, and should enable the experimenter to analyze and evaluate whether lipid messenger systems are likely to play any significant roles in his or her particular model-system. An important experimental point about many lipid signaling systems is that activation of these cascades does not always lead to significant changes in the total cellular content of individual "signaling lipids." Rather, the activation of a lipid signaling cascade becomes manifest through alterations in the rate of turnover (i.e., the net rate of synthesis/degradation) of specific lipid components. It is therefore important to decide whether investigations should be focused on detecting changes in chemical levels of specific lipids or whether the assessment of changes in turnover rate is the aim. In this chapter some methods will be presented for the isolation, separation, detection, and, where appropriate, quantification of potential plant signaling lipids. It is impossible to give an exhaustive treatise of this area in the limited space available, so the methods below should be seen merely as a "starter-pack" for investigators new to the plant lipid-signaling field. The brief bibliography will point to additional sources of information about some of the more specialized approaches currently in use.

1.2. The Phosphoinositide System

Since most of the "classical" features of lipid-based signaling systems in both animals *(5,6)* and plants *(7–11)* have been extensively reviewed, we will just give a very brief overview of one of the best characterized eukaryotic lipid-signaling systems, namely the phosphoinositide system. A simplified and schematic model of the PI-system is shown in **Fig. 1**.

Fig. 1. Generation of the second messenger, inositol(1,4,5)trisphosphate by phospholipase C mediated hydrolysis of the lipid phosphatidylinositol(4,5)bisphosphate. This figure gives a simplified presentation of some of the key reactions leading from receptor occupancy to inositol(1,4,5)trisphosphate generation and mobilization of intracellular

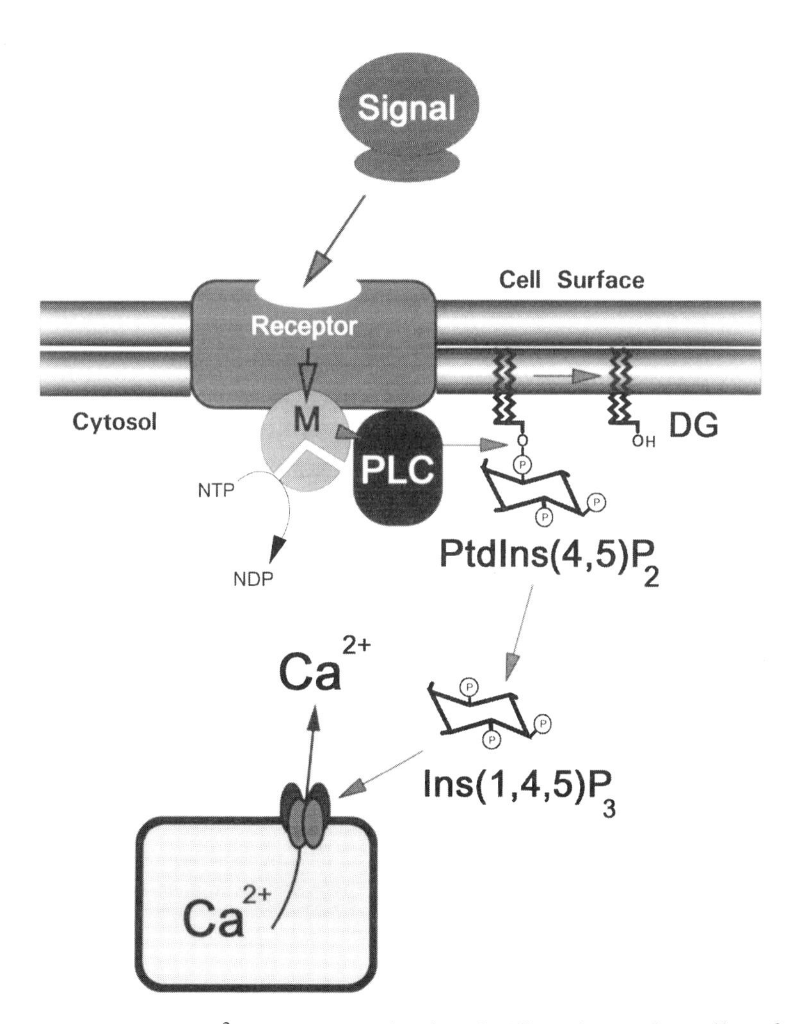

(Fig. 1 continued) Ca^{2+}. When certain signals (S) arrive at the cell surface and associate with specific cell-surface receptors the enzyme phospholipase C (PLC) is activated (often via the interaction with a mediator protein (M). (PtdIns(4,5)P_2), which results in the formation of the two second-messenger molecules: inositol(1,4,5)trisphosphate (Ins(1,4,5)P_3) and diacylglycerol (DG). Ins(1,4,5)P_3 is capable of specifically inducing Ca^{2+}-release from intracellular stores, whereas DG in many cell types modulates the activity of a group of enzymes known collectively as protein kinase(s) C. The concomitant increase in cytosolic Ca^{2+} and the switch-on of protein kinase activity results in a bifurcated signal and cell activation. When the stimulus ceases to exert its effect, the agonist-receptor complex is dissociated and PLC reverts to its inactive configuration. When PtdIns(4,5)P_2 is no longer hydrolyzed the cellular levels of Ins(1,4,5)P_3 and DG decrease and the cytosolic Ca^{2+} concentration return to low-n*M* levels.

In non-stimulated animal and plant cells the free cytosolic Ca^{2+} concentrations are carefully maintained at low nanomolar levels. Since the concentration of Ca^{2+} in the extracellular space and inside most organelles is many orders of magnitude higher than in the cytosol, a very steep Ca^{2+} gradient exists across the plasma membrane and across endomembranes. Maintaining this gradient is achieved by a number of active Ca^{2+} transport systems located both in the plasma membrane and in other organellar membranes. Polyphosphoinositides in the inner leaflet of the plasma membrane (as well as in other subcellular locations) are continuously formed and degraded by a set of specific kinases and phosphatases, and are thus kept in a state of constant turnover. When specific signals (known collectively as agonists) associate with specific cell-surface receptors the enzyme phosphoinositidase C (PIC, also known as phosphoinositide-specific phospholipase C) is activated. Many PIC isoforms exist in mammalian cells, and several isoforms have also been identified in plant cells. Each of the PIC-isoforms appears to have an isoform-specific mode of activation. The mammalian PIC-β-isoforms are thus activated through regulatory, heterotrimeric GTP-binding proteins, whereas the mammalian γ-isoforms depend on receptor-associated tyrosine kinase(s) for activation. The precise mode of activation of the plant PIC-isoform(s) remains unclear but snippets of evidence suggest that the Ca^{2+}-ion itself may play a key role in the regulation of plant PIC activity. The activation of PIC in both mammalian and plant cells leads to the hydrolysis of the lipid phosphatidylinositol(4,5)bisphosphate (PtdIns(4,5)P_2), which results in the liberation of the two second-messenger molecules: inositol(1,4,5)trisphosphate (Ins(1,4,5)P_3) and diacylglycerol (DG). Ins(1,4,5)P_3 is capable of specifically inducing Ca^{2+}-release from intracellular stores, whereas DG in many cell types modulates the activity of a group of enzymes known as protein kinase(s) C (PKC). The concomitant increase in cytosolic Ca^{2+} and the switch-on of protein kinase activity results in a bifurcated signal and cell activation. When the stimulus ceases to exert its effect the agonist-receptor complex is dissociated and PIC is converted back into an inactive configuration. When PtdIns(4,5)P_2 is no longer hydrolyzed the cellular levels of Ins(1,4,5)P_3 and DG decrease and the cytosolic Ca^{2+}

levels return to low n*M* courtesy of the Ca^{2+}-transporting systems removing Ca^{2+} from the cytosol. All the central structural and functional components of the mammalian phosphoinositide system have now been identified in plant cells and the reader is referred to recent reviews for further details *(10–12)*. The precise role(s) of the phosphoinositide system in plant responses to plant hormones (growth factors) is still being debated but changes in inositol lipid metabolism have been reported to accompany responses to auxin, cytokinin, and giberellic acid (for references *see* **ref.** *7*). More recently, data have been presented that suggest that a causal link exists between abscisic acid, phosphoinositide metabolism, and control of stomatal cell aperture *(13)*. Slower and more broad-ranging effects of plant hormones on plant phospholipids have also been documented in the literature *(14)*.

1.3. Glycolipids and Cell Signaling

The role of glycolipids in plant cell signaling and hormone responses is even less well understood than that of the phosphoinositides, and only very few experiments have to date been carried out to investigate this possibility. However, the importance of glycolipids/glycosphingolipids in a range of signaling events, including hormone responses in mammalian and other eukaryotic cells *(4)*, suggests that an investigation of whether similar phenomena may also occur in plant cells is timely and could prove worthwhile. It has already been shown that complex glycolipids are intimately involved in certain developmental processes in plant cells, such as the formation of root nodules *(15,16)*, but until now the lack of appropriate methodology has slowed down progress in this field of research. The new and highly sensitive method for the detection and quantification of glycolipids described in this chapter should help to overcome at least some of these technical obstacles.

2. Materials

2.1. Extraction of Phosphoinositides (Phospholipids) and Complex Glycolipids

1. Quench-solution: ice-cold $CHCl_3$-methanol-conc HCl (100:100:0.7, v/v/v).

2. Wash solution: $CHCl_3$-MeOH-0.6 *N* HCl (3:48:47, v/v/v).
3. Cooled homogenizer (e.g., a Kondes glass-grinder).
4. Stoppered glass centrifuge tubes.
5. Pasteur pipets and small glass storage containers.

2.2. Biotinylation of Complex Glycolipids

1. Soda-glass vials and glass centrifuge tubes.
2. 100 m*M* Na acetate/acetic acid buffer, pH 5.5 (reaction buffer A).
3. Bath-type sonicator.
4. Sodium *meta*periodate (150 µL of 0.5 mg/mL in reaction buffer A).
5. Sodium bisulfite (20 mg/mL in reaction buffer A).
6. 5 m*M* biotin hydrazide in dimethylformamide.
7. Methanol-chloroform (1:1, v/v).

2.3. Separation of Phosphoinositides and Biotinylated Glycolipids

2.3.1. Method I

1. An appropriate Thin Layer Chromatography (TLC) plate (e.g., High Performance Thin Layer Chromatography (HPTLC) plate; 60 A HPK Whatman, UK, although other types of plates can also be used).
2. 1% (w/v) potassium oxalate containing 1.5 mM ethylenediamine tetra-acetic acid (EDTA).
3. Chloroform:methanol (1:1, v/v).
4. A sealed glass TLC tank.
5. Developing solvent: chloroform:methanol:ammonia:water (45:35:2:8, v/v/v/v).
6. Commercial standard mixtures of phospholipids/glycolipids. Authentic phospholipid and glycolipid standards are prepared by dissolving 1 mg lipid in 1 mL of methanol:chloroform (1:1, v/v). A range of commercially available lipids can be used (e.g., Sigma, St. Louis , MO): L-α-phosphatidylethanolamine (PE, di 18:0), L-α-phosphatidylinositol (PI, 18:2 and 16:0), D,L-α-phosphatidyl-L-serine (PS, di 18:0), L-α-phosphatidylcholine (PC, lecithin from soybean enriched in 16:0, 18:1, and 18:2), sphingomyelin (Sph, 18:0), and mixtures of galactosylcerebrosides (GalCer), glucosylcerebrosides (GlcCer), and lactosylcerebrosides (LacCer). Polyphosphoinositides can be obtained either in a mixture with other mammalian lipids from

Sigma (brain extract) or can be bought individually in purer form from Calbiochem (Nottingham, UK) or Matreya, Inc. (Pleasant Gap, USA) All organic solvents must be analytical grade. Organic solvent mixtures should always be freshly prepared immediately before use.

2.3.2. Method II

1. TLC plates and equipment as in **Subheading 2.3.1.**
2. 4.55 g disodium trans-1,2-Diaminocyclohexane-*N,N,N',N'*-tetra-acetic acid (CDTA) dissolved in 165 mL water, 330 mL ethanol, and 3.0 mL NaOH.
3. TLC running solvent: methanol (75 mL), chloroform (60 mL) pyridine (45 mL) and boric acid (12 g), water (75 mL), formic acid (3 mL, 88%, v/v), and butylated hydroxytoluene (BHT, 2,[6]-Di-tert-Butyl-*p*-cresol, 0.375 g).

2.4. Detection and Quantification of (Biotinylated) Glycolipids and Phospholipids (Phosphoinositides) After Separation by TLC or Transfer to PVDF Membranes

2.4.1. Chemical Detection of Glycolipids and Phosphoinositides on TLC Plates

1. 3% (w/v) cupric acetate in 8% H_3PO_4 (v/v).
2. Acid-resistant spray-can.
3. Ventilated oven at 180°C.

2.4.2. Detection of Biotinylated Glycolipids

1. PVDF blotting membrane (Hybond-PVDF, Amersham, UK).
2. Blotting solvent A: propan-2-ol:methanol:0.2% aqueous $CaCl_2$, 40:20:7 (v/v/v).
3. Hot laundry iron (80–100°C).
4. A thin sheet of glass-fiber (GF/A Whatman, UK).
5. Primuline fluorescent reagent (0.5% w/v in water).
6. UV light source.
7. Freshly prepared methanol:water (75:25 v/v).
8. Blocking solution containing 3% BSA (w/v) in phosphate-buffered saline (PBS, 100 m*M* Na_2HPO_4/NaH_2PO_4 and 100 m*M* NaCl, pH 7.5).

9. Streptavidin-horseradish peroxidase (SHRP; Amersham, UK).
10. Enhanced chemiluminescence (ECL) reagent (Amersham, UK).
11. Radiograph film (e.g., Fuji RX, Japan).
12. Dark-room facilities.
13. Scintillation counter.

2.5. Quantification of EASTern Blotted Biotinylated Glycolipids

1. Scanning equipment (e.g., HP ScanJet 3C/ADF with DeskScan II HP software or another appropriate scanning set-up).
2. PC with ImageQuant 3.3 software (or equivalent).

3. Methods

3.1. Extraction of Phosphoinositides (Phospholipids) and Complex Glycolipids

1. Prepare the tissue/cells to be extracted in an appropriate fashion (e.g., radiolabeling, hormone treatment).
2. Remove the incubation medium rapidly by either filtration or aspiration. Immediately transfer the plant tissue into Quench-solution: ice-cold $CHCl_3$-methanol-concentrated HCl (100:100:0.7, v/v/v). The volume of Quench-solution required depends on several factors, including the amount of tissue to be extracted and its moisture content. Approximately 10 mL Quench-medium should be used per 0.5 g fresh tissue. If tissues with low water content are to be extracted lower volumes of Quench-solution can be used.
3. Rapidly transfer the cell/tissue suspension to a cooled homogenizer (e.g., a Kondes glass-grinder) and homogenize the tissue. A one-phase system should be visible after the homogenization. If the tissue has a very high moisture content or if significant amounts of the incubation medium are transferred with the tissue, a two-phase system may form at this stage. If this happens a small volume of cold methanol must be added until a one-phase system is again obtained.
4. 2 mL 0.6 *N* HCl (per 10 mL of Quench-solution) is then added to the tissue homogenate. This creates a two-phase system consisting of an acidified aqueous upper phase and an organic lower phase contain-

ing relatively nonpolar compounds, such as phospholipids and glycolipids (*see* **Note 1**).

5. Carefully remove the top phase with a glass-pipet and discard it (*see* **Note 2**).
6. Wash the remaining bottom phase three times with 4-mL aliquots of $CHCl_3$-MeOH-0.6 *N* HCl (3:48:47, v/v/v), each time removing and discarding the top phase.
7. Transfer the bottom phase and any interface material to stoppered glass centrifuge tubes and centrifuge at 5000*g* for 4–5 min (*see* **Note 3**).
8. Remove most of the top phase and recover the bottom phase with a glass Pasteur pipet. Transfer this to small glass storage containers.
9. Evaporate the lipid extract to dryness under a stream of O_2-free N_2.
10. Add a small volume (e.g., 50–300 µL) $CHCl_3$-MeOH (1:1, v/v) to redissolve the lipids.
11. Flush the storage container briefly with O_2-free N_2 and seal it immediately. Lipid extracts are best stored in the dark at –20°C. With time, oxidation of fatty acids and degradation to lyso- and glycerophosphoryl derivatives does occur. For accurate quantitative experiments the best results will obviously be obtained if the lipid extracts are analyzed (quantified) as soon after preparation as possible (*see* **Note 4**).
12. The lipid extract described above (**steps 1–11**) is suitable for the analysis of both phospholipids (phosphoinositides) and glycolipids. If the extract is to be used for both it is divided into two portions and further processed as described below.

3.2. Biotinylation of Complex Glycolipids

1. Transfer the lipid extract containing glycolipids (25 µL, in methanol:chloroform, 1:1, v/v) to soda-glass vials.
2. Add 225 µL of 100 m*M* Na acetate/acetic acid buffer, pH 5.5 (reaction buffer A) to the lipid extract and sonicate the mixture for 15 s in a bath-type sonicator. Vortex the vial thoroughly and sonicate again for 15 s. This will result in the formation of a lipid suspension/emulsion.
3. Add sodium *meta*periodate (150 µL of 0.5 mg/mL in reaction buffer A) and vortex the vial vigorously for 10–15 s.
4. Cover the vials with aluminum foil to exclude light and allow the mild-oxidation reaction to proceed for 15 min at room temperature. (The *meta*periodate reacts with the carbohydrate portion of the lipids and aldehyde groups are formed) (*see* **Note 5**).

5. Stop the mild-oxidation reaction by adding 150 µL sodium bisulfite (20 mg/mL in reaction buffer A) and incubate the samples for 10 min at room temperature, to eliminate excess *meta*periodate.
6. Start the biotinylation reaction by adding to the sample mixture: 25 µL 5 m*M* biotin hydrazide (Sigma) in dimethylformamide. Incubate the vials for 3 h in the dark at room temperature.
7. Sequentially add 1 mL methanol-chloroform (1:1, v/v) and then 500 µL of chloroform to redissolve the lipid derivatives. Recover the organic phase containing the dissolved lipids and gently evaporate the solvent under O_2-free nitrogen.
8. The biotinylated glycolipids can either be used immediately for further analyses or may be stored either dry at –20°C or as described in **Subheading 3.1.** for the total lipid extracts.

3.3. Separation of Phosphoinositides and Biotinylated Glycolipids

3.3.2. Method I

1. Evaporate any remaining solvents from the phospholipid and/or biotinylated glycolipid samples by evaporation under O_2-free N_2 and redissolve the lipids in 50 µL cold chloroform:methanol (1:1, v/v).
2. Load the lipid samples onto an appropriate TLC plate (e.g., HPTLC plate; 60 A HPK [Whatman, UK], other types of plates can of course also be used) that has been dipped (or sprayed) with 1% (w/v) potassium oxalate and 1.5 m*M* EDTA, and activated for 2 h at 120°C before use. Make sure that the plate has cooled to room temperature before loading the samples.
3. Develop the TLC plate in a sealed TLC tank using the following solvent system: chloroform:methanol:ammonia:water (45:35:2:8, v/v/v/v) (*see* **Note 6**). It is recommended that commercial standard mixtures of phospholipids/glycolipids be used as authentic standards and are developed alongside the unknown lipids as internal standards.
4. Remove the TLC plate from the developing tank when the solvent front has reached 0.5–1.0 cm from the top of the plate. Gently air dry the plate (*see* **Note 7**).

Whereas method I is ideally suited for the separation of most of the common plant phospholipids there is an added degree of complexity with regard to the analysis of the polyphosphoinositides. This is because of the comparatively recent discovery that in addition to the 4-phosphorylated phosphoinositides, PtdIns(4)P and PtdIns(4,5)P_2, several other phosphoinositides also exist that contain a P-monoester in the D-3 position of the *myo*-inositol ring. Since these lipids chemically are identical to the 4-phosphorylated isomers they have identical chromatographic behavior in most commonly used TLC systems. Thus, separation of all known polyphosphoinositide isomers is unlikely to be achieved by simple TLC procedures. However, several procedures for the separation of at least some of these lipids by TLC have been developed. One such method is based on the fact that formation of borate complexes (of different strengths) can occur between hydroxyls in the polyphosphoinositide headgroup and free borate ions. (For further details consult **ref.** ***17***). The method described in **Subheading 3.3.2.** is routinely used in our lab.

3.3.2. Method II

1. Prepare the CDTA wash by stirring a mixture of 4.55 g disodium CDTA*H_2O, 165 mL water, 330 mL ethanol, and 3.0 mL NaOH until CDTA is fully dissolved.
2. Immerse silica gel 60 plates face-up in CDTA wash for 5–10 s.
3. Remove the plate and leave it to dry for 30–60 min.
4. Activate the plate in a ventilated oven (110°C) for 1 h.
5. Prepare the TLC solvent by stirring together 75 mL methanol, 60 mL chloroform, 45 mL pyridine, and 12 g boric acid until the boric acid is completely dissolved (*see* **Note 8**).
6. Add 7.5 mL water, 3.0 mL formic acid (88% v/v) and 0.375 g BHT to the TLC solvent.
7. Apply the lipid samples 1.7 cm from the bottom of the TLC plate.
8. Develop the plate in a thoroughly equilibrated TLC tank and process the plate as described in **Subheading 3.3.1.** (*see* **Note 9**).

The detection of biotinylated glycolipid derivatives and phosphoinositides (phospholipids) is described in the following sections.

3.4. Detection and Quantification of (Biotinylated) Glycolipids and Phospholipids (Phosphoinositides) After Separation by TLC or Transfer to PVDF Membranes

3.4.1. Chemical Detection of Glycolipids and Phospholipids on TLC Plates

Detection of phospholipids and glycolipids can either be carried out directly on the TLC plate or, in the case of glycolipids, after transfer to PVDF membranes. For glycolipids, the latter of the two procedures is by far the more sensitive but does require facilities for general ECL work.

The direct detection of phospholipids (phosphoinositides) and glycolipids on TLC plates can be achieved in a number of ways. A very simple method is described below:

1. Spray the dry plate with 3% (w/v) cupric acetate in 8% H_3PO_4 (v/v).
2. Incubate the plate in a ventilated oven at 180°C for 1 h.
3. Glycolipids (and other lipids) will appear as dark-brown/black spots on a white/yellow background.

Many other spray reagents are available and the reader is referred to standard texts on the subject (e.g., **ref. *18***). Most of these chemical methods, however, are destructive and at best only semiquantitative.

In the case of the plant phosphoinositides (except for phosphatidylinositol) the chemical levels are in most cases so low that it is impossible to detect them using simple chemical detection procedures (such as spray reagents). This has meant that, almost without exception, work on phosphoinositides involves radiolabeling and detection of the separated lipids by autoradiography. Further quantification can then be achieved by recovery of the radiolabeled lipids and determination of incorporated label by liquid scintillation spectrometry. Reliable quantitative information can, of course, only be obtained by this procedure if isotopic equilibrium has been reached. Several precursors can be used for the labeling of phosphoinositides in plant tissues. Most common are ^{32}P (or ^{33}P) orthophosphate and *myo*[2-^{3}H]-inositol, but ^{14}C-glycerol and ^{14}C-acetate have also been used. Having radiolabeled the tissue under study TLC may in some instances give a first indication of whether a dramatic increase in the turnover of polyphosphoinositides has occurred as a result of a specific treatment, but in most cases it is necessary to carry out further, and more detailed investigations before valid conclusions can be reached. It is beyond the scope of this chapter to describe all the methods that can be used for such detailed investigations and it will suffice to say that one of the most common approaches involve the deacylation (O-N transacylation) and deglyceration of the polyphosphoinositides under study followed by high performance liquid chromatography (HPLC) separation of the resulting inositol phosphates and quantification by liquid scintillation spectrometry. Examples of how these procedures have been utilized can be found in **refs.** ***9–11*** and Drøbak and Roberts ***(21)*** have given a step-by-step guide to O-N transacylation of plant phosphoinositides and the separation of the resulting derivatives by HPLC. Even the conversion of inositol lipids to their corresponding inositol phosphate derivatives does not always ensure that adequate information is obtained about the precise individual isomers and further conversion to $^{3}H/(^{14}C)$-labeled alditol derivatives is sometimes required. For further and more detailed information and background to these methods and approaches the reader should consult the two excellent books on the subject edited by Irvine ***(22)*** and Shears ***(23)***, respectively.

3.4.2. Detection of Biotinylated Glycolipids

To allow the detection of HPTLC-separated biotinylated glycolipids using aquous reagents the lipids must first be transferred onto a PVDF blotting membrane (Hyebond-PVDF, Amersham, UK). This procedure: Energy-Assisted-Solid-Transfer (or EASTern blotting) is carried out as follows:

1. Soak the silica plate in blotting solvent A: propan-2-ol:methanol:0.2% aqueous $CaCl_2$, 40:20:7 (v/v/v) for 15 s. Then wet the PVDF membrane for a similar period of time.
2. Transfer the lipids from the TLC plate onto the blot by placing the PVDF blotting membrane on the surface of the TLC plate and

apply a gentle and even pressure to the membrane with a hot laundry iron (80–100°C) for 1 min. It is important to ensure that the PVDF membrane is not brought into direct contact with the iron. To avoid this, insert a thin sheet of glass-fiber (GF/A Whatman, UK) between the iron and the membrane. The successful transfer of samples onto the membrane is monitored by spraying the membrane with primuline fluorescent reagent (0.5% [w/v] in water) followed by visualization of lipids under UV light (*see* **refs. *19*** and ***20***).

3. Prior to the incubation with streptavidin-horseradish peroxidase (SHRP), moisten the blot briefly with methanol:water (75:25 v/v; *see* **Note 10**), and then incubate the blot in a blocking solution containing 3% BSA (w/v) in phosphate-buffered saline (PBS, 100 m*M* Na_2HPO_4/NaH_2PO_4 and 100 m*M* NaCl, pH 7.5) for 2 h at room temperature or overnight at 4°C.
4. Rinse the blot thoroughly with PBS (at least 6 rinses of 75 mL for 30 min).
5. Incubate the membrane with SHRP (Amersham, UK) diluted 1/2000 in PBS for 3 h at room temperature.
6. Rinse the membrane thoroughly with PBS as described in **step 4** above.
7. Prepare the ECL reagent (Amersham, UK) from solutions designated "A" and "B" by the manufacturer by mixing them in the ratio 1:1. Dilute the ECL reagent fivefold in deionized water (*see* **Note 11**).
8. Add the ECL reagent to the PVDF membrane and expose the blot to radiography film (Fuji RX, Japan) for an appropriate length of time (*see* **Notes 12** and **13**).

Figure 2 shows a typical ECL-film of glycolipids separated, EASTern blotted, and detected as described in **Subheading 3.4.2.**

3.4.3. Quantification of EASTern-Blotted Biotinylated Glycolipids

When the procedure outlined in **Subheading 3.4.1.** has been followed the glycolipids can be quantified by scanning the ECL-films using, e.g., a HP ScanJet 3C/ADF with DeskScan II HP software or another appropriate scanning set-up. **Figure 3** shows the optical density of peak areas (given as scanner units, i.e., number of pixels) quantified by ImagQuant 3.3 software.

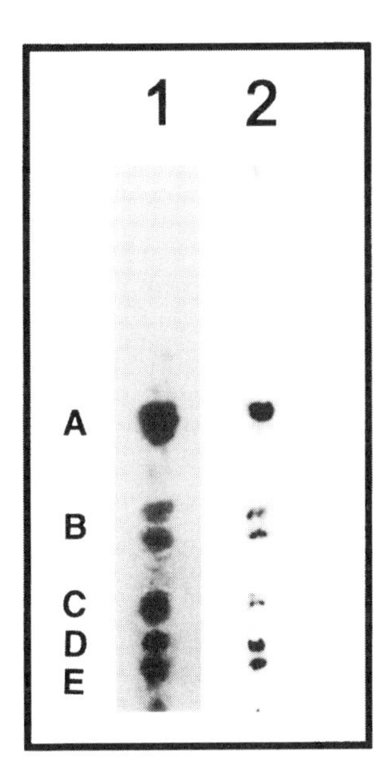

Fig. 2. Detection of glycolipids (5 µg per lane) using **(A)** the ECL procedure described in **Subheading 3.4.2.** and **(B)** the Cu-acetate staining procedure outlined in **Subheading 3.4.1**. The glycolipids are: galactosylcerebroside, Galβ1-1Cer; lactylceramide, Galβ1-4Glcβ1-1Cer; trihexosylceramide, Galα1-4Galβ1-4Glcβ1-1Cer; globoside, GalNAcβ1-3Galα1-4Galβ1-4Glcβ1-1Cer' Forssman glycolipid, GalNAcα1-3GalNAcβ1-3Galα1-4Galβ1-4Glcβ1-1Cer.

4. Notes

1. It is very important for the extraction of polyphosphoinositides that an acidic environment's maintained throughout the extraction procedure. The presence of phosphomonoester groups in these lipids makes them quite polar despite their lipid nature. If neutral or near-neutral extraction systems are used a significant loss of polar lipids to the aqueous top phase is likely to occur.
2. If water-solube radioactive precursors (such as ^{32}P) have been used for lipid labeling a large proportion of the unincorporated label will be present in the top-phase of the extraction system and particular care should be taken to avoid contamination or exposure.
3. It is recommended that the glass tubes be closed with, e.g., silicone bungs to avoid evaporation and exposure to organic vapors.

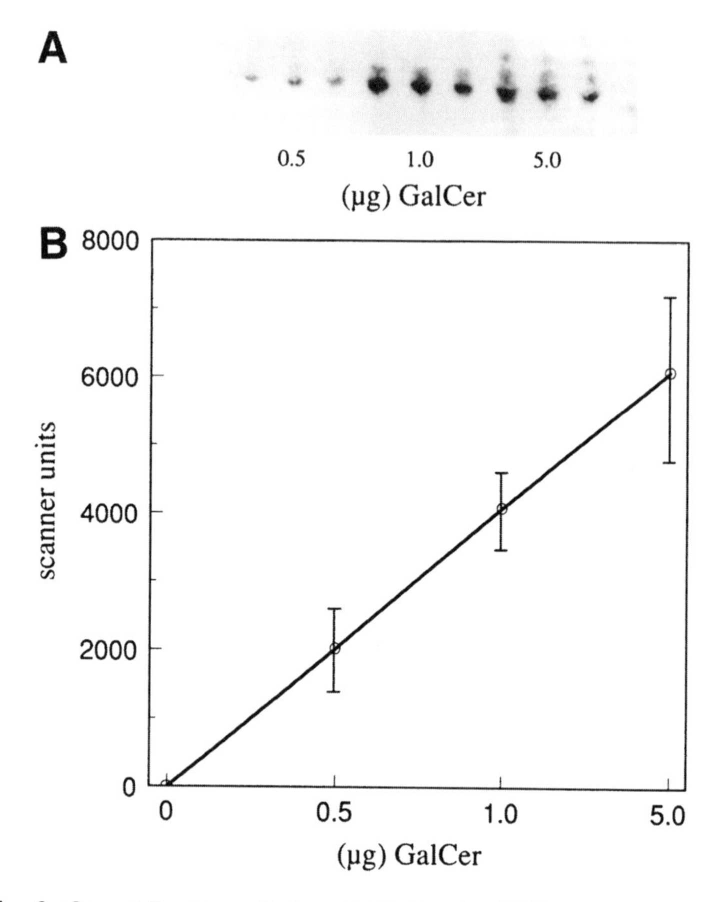

Fig. 3. Quantification of glycolipids by the ECL system. **(A)** Radiograph of three replicates of GalCer (concentrations: 0.5, 1.0, and 5.0 μg) and detected with ECL (1:1:10, reagent A, B and H_2O exposed for 1.5 min). **(B)** Standard curve of mean optical density vs GalCer concentrations. Bar indicates standard error of 15 replicates ($\gamma^2 = 0.99$, $p < 0.05$).

4. If significant amounts of unsaturated fatty acids are suspected to be present in the lipid extract it is recommended that 0.05% butylated hydroxytoluene is added as an antioxidant.
5. Aldehyde groups react spontaneously with biotin hydrazide.
6. The running solvent should always be made up fresh shortly before use. Ensure that the atmosphere in the TLC tank has thoroughly equilibrated with the running solvent before the TLC plate is inserted.
7. If rapid drying is required a commercial hair-drier can be used.

8. Ensure that evaporation does not occur during the preparation of the running solvent.
9. The pyridine smell can take a considerable time to disappear so the TLC plate should be kept in a well-ventilated place (e.g., fume hood) until all solvents have evaporated completely.
10. Prolongated exposure of the blot to methanol should be avoided because some lipids may be dissolved and lost from the membrane.
11. The right dilution of the reagent can be worked out relatively easily as ECL allows the reprobing of the SHRP. Thus, it is simply a matter of trial and error to adjust the concentration of the reagents to individual needs.
12. Excess ECL-reagent during exposure can cause blurry images. Wrap the blot in, e.g., cling film to ensure that an even and dry surface is exposed to the autoradiograph film.
13. The optimal exposure time can be determined by trial and error.

References

1. Michell, R. H. (1975) Inositol phospholipids and cell surface receptor function. *Biochim. Biophys. Acta* **415,** 81–147.
2. Berridge, M. J. and Irvine, R. F. (1989) Inositol phosphates and cell signalling. *Nature* **341,** 197–205.
3. Nishizuka, Y. (1992) Intracellular signaling by hydrolysis of phospholipids and the activation of protein kinase C. *Science* **258,** 607–614.
4. Hakomori, S.-I. (1993) Structure and function of sphingoglycolipids in transmembrane signalling and cell–cell interactions. *Biochem. Soc. Transactions* **21,** 583–595.
5. Berridge, M. J. (1993) Inositol trisphosphate and calcium signalling. *Nature* **361,** 315–325.
6. Irvine, R. F. (1992) Inositol lipids in cell signalling. *Curr. Opin. Cell Biol.* **4,** 212–219.
7. Drøbak, B. K. (1992) The plant phosphoinositide system. *Biochem. J.* **288,** 697–712.
8. Cote, G. G. and Crain, R. C. (1993) Biochemistry of phosphoinositides. *Ann. Rev. Plant Physiol. Plant Mol. Biol.* **44,** 333–356.
9. Drøbak, B. K. (1996) Metabolism of plant phosphoinositides and other inositol-containing lipids, in *Membranes: Specialized Functions in Plants* (Smallwood, M., Knox, J. P., Bowles, D. J., eds.), BIOS Scientific Publishers, Oxford, UK, pp. 195–214.

10. Munnik, T., Irvine, R. F., and Musgrave, A. (1998) Phospholipid signalling in plants. *Biochem. Biophys. Acta* **1389**, 222–272.
11. Mazlik, P. (1997) Plant lipids in cell signalling: a review. *Ann. Biol.* **36,** 165–180.
12. Drøbak, B. K., Dewey, R. E., and Boss, W. F. (1999) Phosphoinositide kinases and the synthesis of polyphosphoinositides in higher plant cells. *International Review in Cytology—A survey of Cell Biology* **189,** 95–130.
13. Lee, Y. S., Choi, Y. B., Suh, S., Lee, J., Assmann, S. M., Joe, C. O., Kelleher, J. F., and Crain, R. C. (1996) Abscisic acid-induced phosphoinositide turnover in guard-cell protoplasts of *Vicia faba. Plant Physiol.* **110,** 987–996.
14. Mudd, J. B. (1980) Phospholipids biosynthesis, in *The Biochemistry of Plants—A Comprehensive Treatise,* vol. 4. (Stumpf, P. K., ed.), Academic Press, New York, pp. 249–282.
15. Perotto, S., Donovan, N., Drøbak, B. K., and Brewin, N. J. (1995) Differential expression of a glycosyl inositol phospholipid antigen on the peribacteroid membrane during pea nodule development. *Mol. Plant Microbe Interac.* **8,** 560–568.
16. Hernández, L. E., Perotto, S., Brewin, N. J., and Drøbak, B. K. (1995) A novel inositol-lipid in plant-bacteria symbiosis. *Biochem. Soc. Trans.* **23,** 582S.
17. Walsh, J. P. Caldwell, K. K., and Majerus, P. W. (1991) Formation of phosphatidylinositol 3-phosphate by isomerization from phosphatidylinositol 4-phosphate. *Proc. Natl. Acad. Sci. USA* **88,** 9184–9187.
18. Christie, W. W. (1982) *Lipid Analysis.* Pergamon, Oxford, UK.
19. Taki, T., Handa, S., and Ishikawa, D. (1994) Blotting of glycolipids and phospholipids from a high-performance thin-layer chromatogram to a polyvinylidene difluoride membrane. *Anal. Biochem.* **221,** 312–316.
20. Taki, T., Kasama, T., Handa, S., and Ishikawa, D. (1994) A simple and quantitative purification of glycosphingolipids and phospholipids by thin-layer chromatography blotting. *Anal. Biochem.* **223,** 232–238.
21. Drøbak, B. K. and Roberts, K. (1992) Analysis of components of the plant phosphoinositide system in *Molecular Plant Pathology: A Practical Approach*, vol. 2. (Gurr, S. J., McPherson, M. J., and Bowles, D. J., eds.), IRL, Oxford, UK, pp. 195–221.
22. Irvine, R. F. (1990) *Methods in Inositide Research.* Raven, New York.
23. Shears, S. B. (1997) *Signaling by Inositides—A Practical Approach.* IRL, Oxford, UK.

12

Reverse Genetics

Screening Plant Populations for Gene Knockouts

Sean T. May, Deborah Clements, and Malcolm J. Bennett

1. Introduction

1.1. Reverse Genetics

Functional genomics has recently become a logistical rather than a theoretical problem in plant molecular biology for one simple reason—an immense wealth of data describing the nucleotide sequences of plant and animal genomes has become available and is continuing to accrue at an exponential rate. This has led to the development of a variety of methods under the general heading of functional genomics through which this ocean of information can be charted and given biological meaning with regard to the significance of individual nucleotide sequences. In order to assess the biological impact of individual genes within the context of a living organism, one might wish to investigate mutations or lesions in the gene of choice. Unfortunately, mutated lines are normally derived following selection based on an observation of phenotype, which then leads to an investigation of the molecular genetic changes underpinning the phenotypic changes observed. This forward genetic approach is very powerful, but does not lend itself readily to the investigation of genes without a known function (such as the analysis of theoretically pre-

From: *Methods in Molecular Biology, vol. 141: Plant Hormone Protocols*
Edited by: G. A. Tucker and J. A. Roberts © Humana Press Inc., Totowa, NJ

dicted transcripts derived from genome sequencing projects). In order to design smart screens to isolate suitable lesions, one may require a large degree of prior knowledge or guesswork relating to the suspected function of that gene, or else be driven toward screens that are essentially directed through known phenotypes.

The essence of reverse genetics is that one could take a nucleotide sequence; for example, from a genomic or transcriptomic (e.g., expressed sequence tag–EST) database *(1)*, and derive or identify individuals that harbor lesions in that particular gene. This then allows investigators to select a suitable gauntlet of phenotypic tests in order to approach a derivation of function for their chosen gene. The most efficient route for reverse genetics, a directed insertional mutagenesis method (DIMM), whereby one may specifically determine a target in the genome and cause lesions to appear under the absolute control of the investigator *(2)*, has not currently been brought to reliable maturity in plant systems. A widespread and viable alternative, however, has been used in a range of eukaryote systems, including plants, and will be described here. This is the selected insertional mutagenesis method (SIMM).

The principle of SIMM reverse genetics is most clearly visualized by keeping in mind the desired end-product of the process. Ultimately, one requires an insertion located within or adjacent to a gene of interest such that the gene is disrupted and its transcripts/downstream products are thereby reduced, rendered nonfunctional, or, preferably eliminated. In a single plant with an insert of this desired type (**Fig 1.**), a polymerase chain reaction (PCR) amplification performed on DNA extracted from this plant will produce a diagnostic band from PCR with one primer in the gene directed toward the insert (primer g), and the other in the insert directed toward the gene (primer i).

Exactly the same diagnostic band will be produced when PCR is performed on DNA extracted from a population that includes one "hidden" plant carrying the required insert among a large number of other plants which are wild-type for the gene of interest. The sensitivity of PCR amplification easily allows the detection of only one positive plant with a specific insertion among a population of more than a thousand wild-type negatives.

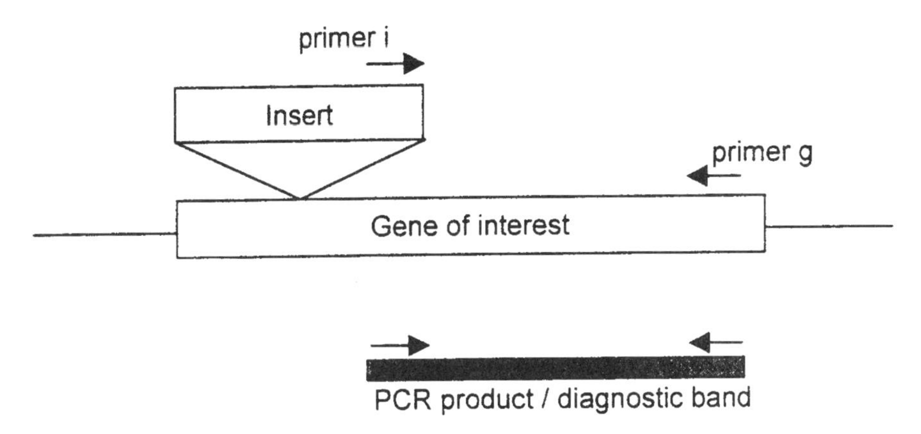

Fig. 1. Detection of an insertion within a given gene by PCR amplification.

The power of PCR-selective reverse genetics is therefore that a population of thousands of plants that have been subject to insertional mutagenesis can be screened for inserts in and near regions of known sequence. To identify the particular plant of interest within that population, one can progressively focus on smaller and smaller positive subpools until the individual plant is identified. For example, a population of 10,000 plants can be split into 10 pools, each containing 1000 plants. A positive pool can then be subdivided further into pools of 100, then subdivided into pools of 10, until finally a single positive plant can be identified that carries the required insertion. Each time, only 10 PCRs are required to identify the positive pool so that the total number of PCRs is only 10×4 (40) instead of 10^4 (10,000). In a three-dimensional grid procedure *(3)*, the same logic is apparent, except that 30 of the PCRs are done simultaneously instead of successively. Ten PCRs are performed on pools of 1000 and a positive pool is isolated. One can then organize the 1000 plants in a $10 \times 10 \times 10$ three-dimensional array (e.g., column, row, tray) and mix all 10 DNAs from each ordinate of these axes in turn to get 30 PCRs, i.e., all the C1s, C2s .... C10's together, then all the R1s, R2s, and so forth. This approach saves a great deal on time, but requires greater organization and can be difficult to appreciate.

Robotics can be used to dramatically facilitate the repetitive analysis of large insertional populations. In the case of the SLAT filters (Sainsbury Laboratory Arabidopsis Transposants; John Innes Centre, Norwich, UK), nylon filter-bound arrays have been produced that currently carry the representations of more than 40,000 transposon insertions in *Arabidopsis* presented in an array of 864 spots. Each spot holds PCR amplifications of the local flanking regions for 50 transposant lines. By hybridizing suitable probes to these macroarrays, one can leap forward to a pool of 50 lines that may be subsequently analyzed to identify single plants using a reverse genetic approach. With the advent of improved macro- and micro arraying robotics technology, future analysis of large populations should become increasing accessible to the plant science community through service provision of these types of arrays.

1.2. Inserts

This technique only requires knowledge of one or more small regions of nucleotide sequence, which are known for both the gene of interest and for the chosen insertional mutagen. The commonly used insertional mutagens in plant systems fall into two major groups: transposons, such as Ac/Ds and En; and *Agrobacterium* mediated T-DNA insertions ***(4)***. Possible alternative insertional mutagens, delivered by viral infection, for example, could be approached in a similar reverse genetic manner.

Primarily it is essential to identify a region of insert that is reliably and wholly transferred from the delivery system to the target genome. In the case of T-DNAs this can be the well-characterized right and left borders (RB/LB; **5**), for many transposons it can commonly be regions immediately internal to the site of recognition for the transposase. Regions should preferably be chosen that are close to the predicted border or junction where the inserted DNA meets its new genomic context. This is because PCR is most effective and efficient when amplifying short stretches of DNA, up to approx 3 kB (although *see* **Note 1**). DNA between the insertion primer site (primer i; **Fig.1**) and the insert/host genome junction is largely wasted space (although *see* **Note 2**), and will reduce the effective range of detection for the second gene-specific primer (primer g; **Fig. 1**). In some cases, the junction may not be rigorously defined (*see* **Note 3**), so judgement based on the likely inclusion of the sequence will have to be exercised.

1.3. Populations

The choice of insert will largely be determined by preference or more reasonably by access to the population that is to be screened. Several of the initial *Arabidopsis* reverse genetics screens *(6,7)* were performed on the ~6000 lines of T-DNA insertions developed in Arizona *(8)*. The material to reproduce this population has been made available both as seed and DNA pools through the international *Arabidopsis* stock centers *(9)*. In addition, at the time of writing at least two other large populations of 5000+ T-DNA lines donated by Dr. Thomas Jack (US) and by Bechtold & Pelletier (INRA Versailles) are available from the stock centers *(9)*. It is clear that additional large populations using both transposon and T-DNA based mutagens are becoming generally available not only for *Arabidopsis* but increasingly in a number of important species, including crop species, such as maize, rice, and *Brassicas*. The initial step in designing a screening strategy is therefore to ascertain the availability of suitable populations. Suitability may depend on a small number of criteria; the most prominent of which is access to the population or alternatively the expense of time and effort in creating your own population. Second, the degree to which the population is representative may be important (representation is the probability that any given gene has been "hit" in the population under consideration; *see* **Note 4**). Third, if a population is to be created it is important to consider the time and effort involved in growing plants and/or acquiring material or DNA samples for each of the subdivision stages and the organization and storage of these resources.

One partial alternative to mass population screens can be used in some genetically mature plant systems, such as *Arabidopsis*. Using transposons that have a preference for "local jumps," it is now a viable option to generate small populations that are intensively mutagenized for a defined area of the genome. An immobile transposon, at a defined location close to a mapped gene of interest, can be induced to "jump" locally by the temporary introduction of a source of transposase. The next or subsequent generations can then be screened using a reverse genetics methodology by subdivision as usual, but on a much smaller scale. Availability of suitable insertion populations and/or mapped transposons is rapidly becoming easier and more extensive and should not constitute a major problem for most laboratories.

1.4. Genes and Gene Primers

Reverse genetics may be used to find knockouts in specific genes based on very little sequence information, such as may be readily available for novel ESTs. It is therefore suitable for finding insertions in genes for which there is no identified function except for that presumed by homology to previously characterized genes. It may similarly be used to identify lesions in known genes and in genes with sequence sufficiently homologous to known genes such that they can bind suitably designed primers. Our experience has been that in the Feldmann *Arabidopsis* population of approx 10,000 insertions, we tend to find insertions in about one-third to one-fourth of the genes that we search for. It is therefore often preferable to search for members of a conserved family in order to maximize the chances of identifying a suitable hit.

Primers should be positioned to maximize the chances of finding a hit inside the coding region or promoter of the chosen gene. Obviously, the position of the primer will be strongly influenced by considerations of homology when investigating gene families and will also be constrained by standard nucleotide sequence requirements for primer design as detailed below. However, in our opinion, it is optimal to design reverse-complement primers as close to the 3' end of the gene as possible in order to cover the maximum target area of the gene. In addition, where minimal information is available about the gene structure or when the coding region and/or introns are known to be large (>3 kB), a second primer should be designed that lies relatively centrally in the gene in order to more reasonably cover the entire length of the gene. In some cases we also recommend designing primers at the 5' end of the gene directed toward the 3' end. This may increase the chances of identifying inserts downstream of this point where the suitable border primer sites in the insert that can interact with 3' positioned primers are missing or incomplete.

1.5. General Primer Design

Primers are subject to the normal considerations applicable to PCR in that they should be checked for palindromes and for the likelihood of binding to other copies of themselves and to each other. To check this, compare all of the primer sequences to each other as reverse complements. Remember that the molarity of the primers in the reaction is very high and that annealing is a probabilistic event dependent on concentration as well as temperature. Consequently, primers that cannot bind to template under PCR conditions can often be seen to bind to each other resulting in the diversion of the efficiency of the reaction into low-mol-wt "primer–dimer" artefacts. As a rule of thumb, perfectly annealing sequences of more than six nucleotides between primers, particularly at the 3' end, should be avoided. Similarly, since many intron regions are T rich ***(10)***, and GC pairs are stronger binding, we recommend that the primers contain approximately equal proportions of all four nucleotides, avoiding stretches of any single nucleotide. Because the PCR template in reverse genetics is a very complex mixture comprising a rich source of uninteresting insert primer sites as well as potentially the target of interest, this diverts a great deal of the efficiency of the reaction toward unwanted amplification. In order to counter this, the gene-specific primers need to be as efficient as possible. Short primers designed previously for sequencing or diagnostic PCRs frequently waste more time than they save. Suitable primers should preferably anneal at an estimated temperature of 65–70°C (*see* **Note 5**). Degenerate primers constitute an obvious additional problem in that increasing the level of degeneracy reduces specificity and consequently increases the potential for mispriming. Additionally, the ratio of perfectly annealing primers is minimally halved every time degeneracy occurs, thereby compromising the efficiency of the reaction. Although degenerate primers are useful, the above points should be kept in mind so that degeneracy is kept to an absolute minimum.

1.6. Confirmation of Results by Hybridization

Although the logic of the screening as outlined above is straightforward, the analysis of the results can be problematic. Chiefly this is because of both the specificity of the PCR and the complexity of the template. If the PCR is running at suboptimal specificity, then a large number of spurious bands can appear from the amplification. Most are random amplification of polymorphic DNA (RAPD)-like artefacts caused by either, or both, of the primers binding such that they produce a nonspecific amplification product. Bands that are consistent between lanes can normally be avoided by optimizing the PCRs before starting (*see* **Subheading 3.3.2.**). Some RAPDs, however, may be formed using one site in the insert and a second nonspecific site in the host genome. Southern hybridization using all or part of the target gene as a probe against the PCR amplification can identify the required specific targets from among this mass of unwanted products. In an identical fashion, amplification products that are derived from complex insert topology may be eliminated (*see* **Note 6**).

2. Materials

2.1. Screening

Materials will depend on the population chosen (**Subheading 1.3.** and **3.1.**)

2.2. DNA

1. Pre-prepared DNA from pooled insert population or fresh, or liquid nitrogen flash-frozen tissue (*see* **Note 7**) from plant pools.

2. Mortar and pestle.
3. Plant DNA purification columns (Qiagen, UK or equivalent), or 30-mL centrifuge tubes.
4. Extraction buffer: 100 m*M* Tris-HCl, pH 8.0; 500 m*M* ethylene diamine tetra-acetic acid (EDTA); 500 m*M* NaCl; 10 m*M* β-mercaptoethanol.
5. Propan-2-ol.
6. 20% Sodium dodecyl sulfate (SDS).
7. TE buffer: 10 m*M* Tris-HCl, pH 8.0; 1 m*M* EDTA.
8. 5 *M* Potassium acetate (KAc).

2.3. PCR Amplification

1. Mineral or paraffin oil.
2. 0.5 mL Centrifuge tubes, fresh pipet tips (*see* **Note 8**)
3. Thermostable DNA polymerase; with associated 10X buffer: e.g., Taq Polymerase (Gibco UK); 200 m*M* Tris-HCl, pH 8.4, 500 m*M* KCl, 15 m*M* $MgCl_2$; (*see* **Note 9**).
4. 2 m*M* dNTP mix (2 m*M* for each of dATP, dCTP, dGTP, and dTTP).
5. 2 μ*M* each specific gene primers (*see* **Note 10**)
6. 2 μ*M* each border primer (**Table 1**).

2.4. Gel Analysis and Hybridization

1. Standard materials for gel electrophoresis and capilliary Southern blotting onto nylon membranes *(17)*.
2. dig-11-dUTP alkali-labile (Boehringer Mannheim [Roche Group, Switzerland]) PCR materials as in **Subheading 2.3.**
3. Nonradioactive digoxigenin chemiluminescent detection kit. (Boehringer Mannheim [Roche Group]) and appropriate solutions described by this kit.
4. Polythene bag roll suitable for sealing/bag sealer.
5. Sandwich boxes suitable for the size of filters produced from the gel.

3. Methods

3.1. Screening

The screening procedure that you follow will depend on the type and size of the population that you are using. Many populations are

Table 1
T-DNA Primers and Positive Control Primers

Left border primer (LB):	5' GAT GCA CTC GAA ATC AGC CAA TTT TAG AC 3' *(3,7)*
Right border primer (RB):	5' GCT CAT GAT CAG ATT GTC GTT TCC CGC CTT 3' *(3)*
Right border primer (RB2):	5' TCC TTC AAT CGT TGC GGT TCT GTC AGT TC 3' *(7)*
Sec Y control primer 1:	5' TTA ATG ACT TCA GCA CGG AAT GGG 3'
Sec Y control primer 2:	5' CAT TCT TCA AAG CAG ATA TGC CGG 3'

distributed in a defined structure, and should be screened as advised. If, however, you are screening a population or subpopulation that you have produced locally, it is advisable to grid the population into at least a two-dimensional matrix such that you can perform an initial screen in as few PCRs as possible. For example, it is reasonable to perform a PCR on 1000 individuals as a pool and expect a positive PCR band detection if a positive plant is present.

Remember that these 1000 pools would, in an ideal world, each represent individual lines with stable inserts. However, in most cases, especially with segregating populations, you will have to calculate the number of individuals for your population that you will have to screen from a given line or seed pool to reasonably expect representation of the total diversity. The suggested sensitivity of 1 in 1000 should reflect this calculation.

1. Divide the population into lots of 1000 (or 1 lot, if the total is fewer than 1000) and treat each lot as follows: Germinate the seeds on soil (or on plates, although this is unlikely to be necessary unless there is a need to select for a biocide). Grid out the plants into convenient receptacles at a stage convenient for repotting; this will depend on the density of sowing and hence the size of the population. A grid of 20 × 50 is convenient for 1000 plants, but whatever the grid, take advantage of making both axes as large as is convenient (*see* **Note 19**).
2. In the first instance, collect tissue from each individual and pool them all together into the 1000 pool lots described in **Subheading 3.1.1.** (the whole population at once if there are fewer than 1000 individuals). Perform a diagnostic PCR or PCRs and hybridization as described in **Subheadings 3.3.** and **3.4.**, on DNA extracted from these large 1000 pool lots. If there are no positives, then this population may be considered not to have your insert of choice.

3. Only after testing thc large pool(s) as above should you proceed to collect tissue from individual plants and pool these into their respective rows and columns (i.e., each plant provides tissue for one row pool and one column pool). These pools can then be processed as above. By this iterative method you should arrive at a single plant or plant(s) with your insert of choice.

3.2. DNA Preparation

1. Dilute the preprepared DNA to 50 ng/µg or grind 1 g of the mixed tissue with liquid nitrogen using the mortar and pestle to a fine powder.
2. Perform Qiagen plant DNA purification columns or equivalent used as instructed, followed by dilution to 50 ng/µg or DNA extraction (*see* **Note 12**):
   a. Transfer powdered tissue to a 50-mL tube and add 15 mL of extraction buffer.
   b. Add 1 mL 20% SDS and shake.
   c. Incubate at 65°C for 10 min with occasional agitation.
   d. Add 5 mL of 5 *M* KAc, mix, and incubate on ice for 20 min.
   e. Centrifuge at 20,000*g* for 15 min.
   f. Decant carefully into a fresh 50-mL tube, avoiding precipitate.
   g. Add 1/2 vol propan-2-ol (10 mL), mix, and store in 20°C freezer for 1 h.
   h. Centrifuge at 20,000*g* for 15 min and then remove all of the supernatant.
   i. Air-dry the pellet at room temperature for 20 min.
   j. Resuspend the pellet in 500 µL TE prewarmed to 60°C.

3.3. PCR Amplification

3.3.1. General PCR Conditions

All PCRs are performed in a minimum of 10 µL total volume (*see* **Note 13**).

1. The general PCR megamix (*see* **Note 14**) is calculated as follows:

Stock	Vol 100 PCR nopri megamix	Final concentration
2 mM dNTPs	(100 µL)	200 µM
10X buffer	(100 µL)	1X buffer
ddH_20	(650 µL)	

2. Multiple 8.5-µL aliquots of this nopri megamix can be used for each PCR with the addition of 0.5 µL each 2 µ*M* primer and 0.4 µL DNA template.
3. The PCR tube is then covered with one drop of oil from a 200 µL pipeter (*see* **Note 13**) before performing a hot start (*see* **step 6**).
4. Multiple PCRs that require identical primer pairs (such as the screening set) should have each appropriate primer added (*see* **Note 13**):

Stock	Vol 50 PCR +pri megamix	Final concentration
Nopri megamix	425 µL	
2 µ*M* primer 1	50 µL	200 n*M*
2 µ*M* primer 2	50 µL	200 n*M*

5. Multiple 9.5-µL aliquots of this +pri megamix can be used for PCR after the addition of 0.4 µL DNA template. Then they are covered in **step 3**.
6. PCR cycles are performed using the following parameters: 95°C 2 min, "hot start", then addition of 0.1 µL (0.5 U) thermostable polymerase at 72°C (*see* **Note 15**); followed by 35 cycles of 15 s at 94°C; 15 s at annealing temperature (*see* **Note 5**); and 60 s at 72°C (*see* **Note 16**), 1 cycle of 5 min at 72°C to ensure complete extension of all products.

3.3.2. Controls

3.3.2.1. DNA Quality

It is important to test all of the DNA samples that you have prepared individually for their quality and "PCRability." We routinely perform a test PCR on every DNA sample, whether fresh or archival, with a primer pair designed to a known gene *SecY* (*see* **Note 10**; **Table 1**).

3.3.2.2. Gene-Specific Primer Quality

Test your gene-specific primers in order to optimize your annealing temperatures. The annealing temperature chosen from these tests should produce good strong and specific amplification products for pairs that are directed toward each other. The primers should preferably be tested on genomic DNA containing an insert of the type selected for mutagenesis. They should be tested both singly and together and should not produce any unanticipated and therefore nonspecific bands.

3.3.2.3. Insert Specific Primer Quality

In the first instance of use, your insert-specific primers should be tested on wild-type DNA that does not contain any of the mutagenesis inserts. They should be tested at the temperature that you have optimized for the specific primers, and should not produce any amplification products. You should not use DNA that comes from the mutagenized population, particularly for T-DNA populations (*see* **Note 6**).

3.4. Gel Analysis and Hybridization

Nonradioactive digoxigenin based hybridization is the detection method of choice for identifying positive bands (*see* **Note 20**). Protocols can be obtained directly from Boehringer Mannheim (now Roche Group) by post or at: http://biochem.boehringer-Mannheim.com/prod_inf/manuals/dig_man/dig_toc.htm]. It is also possible to use radiolabeled probes ***(17)***, but we do not recommend this.

3.4.1. Gel Analysis

PCR products should be standardly run on 1% agarose gels, acid hydrolyzed, denatured, neutralized, and capilliary Southern blotted onto nylon hybridization filters. The gels will often be very messy to look at and no analysis of possible positives should be made at this stage (*see* **Note 6**). We recommend the use of mini gels (*see* **Note 21**).

3.4.2. Probe

The probe should include as much of the gene of interest as possible, including introns, and can be synthesized by incorporation of alkali-labile Dig-11-dUTP incorporation in a PCR. We can recommend PCR from genomic DNA as long as this produces a clear specific product that can be identified as the anticipated gene. Identification can take the form of simple size identification, nested PCR, or restriction analysis of the fragment obtained.

The protocol for incorporation of dig-11-dUTP label into PCR products is available from Boehringer Mannheim. Simple addition of 0.2 μL of the dig-11-dUTP label (0.2 nmol) into a standard 10-μL PCR normally delivers good probes. However, the yield of the PCR can sometimes be reduced in comparison to PCRs that do not contain the label.

3.4.3. Hybridization and Detection

The protocol for hybridization and detection is available from Boehringer-Mannheim and we strongly advise the use of their Easyhyb hybridization medium for both the prehyb and the actual hybridization. We have found the following specifics to be particularly relevant for our particular use in reverse genetics with *Arabidopsis*:

1. Prehyb and hybridization can be efficiently performed in bags (for minimal volume and economy) placed flat in the bottom of sandwich boxes at 37°C overnight in standard bacterial growth-shaking incubators.
2. We decant the hybridization probe to 25-mL screw-topped "universal" tubes that are stored in the –20°C freezer for long-term storage, and wash the filters to a stringency of 0.1X SSC and 0.1% SDS at 65°C in sandwich boxes.
3. Rinse filters in maleic acid buffer in sandwich boxes, but block and antibody incubate for 30 min each in sealed bags.
4. Wash with maleic acid twice in sandwich boxes (the addition of Tween-20 has not seemed to make any difference to our results); extending these washes by an hour or more can sometimes give cleaner results.
5. Equilibration in sandwich boxes, addition of the flurofor Disodium 3-(4-methoxyspiro{ 1,2-dioxetane-3,2'-(5'-chloro) tricyclo [3,3,1,1 ^{3,7}] decan}-4-yl) phenyl phosphate (CSPD) in bags, incubation at 37°C and then removal of the CSPD, is as directed by the manual. We strongly endorse removal of the excess CSPD as directed. Leaving CSPD on the filter during incubation does not improve the signal, and can lead to unsightly background. CSPD should be carefully squeezed out of the bag before exposure to film.
6. Exposure normally takes about 20 min for a signal.

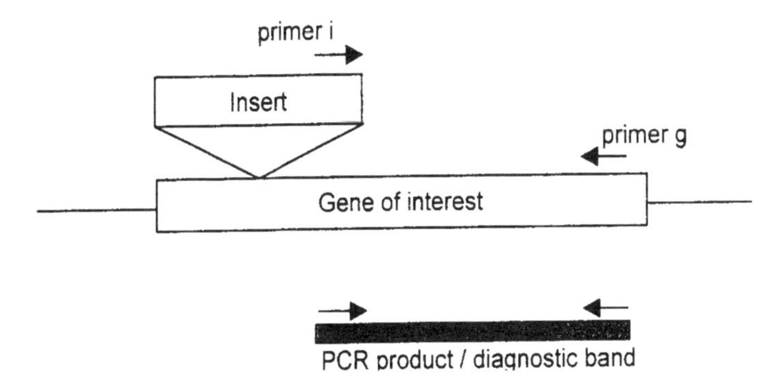

Fig. 2. Shows an example of an actual reverse genetics screen on pools of 100 mixed plants. Lanes 1–4 show a repeated RAPD-like band at approx 400 bp, which is not positively hybridizing. Lane 9 shows a candidate [2] positively hybridizing pool that is not visible on the gel, and lane 1 shows a less effectively hybridizing PCR product [1] showing an insert in a homologous gene to the sequence of interest. The blot has been simultaneously probed with a vector fragment that locates some of the size marker lanes for reference .

3.5. Final Analysis

The results that you get from your gels may be confusing. It is possible that you will get common bands appearing in many lanes, particularly in T-DNA populations. This is caused by RAPD-like nonspecific annealing effects (*see* **Note 6**). In some cases, even the hybridization may be confusing because the probe may bind simply because of the high levels of DNA in these bands. Increasing or extending stringency washes can reduce this. Good positives should be unique in size (unless your population contains multiple iterations of the same insert), and are likely to show high intensity of signal relative to the visible quantity of the band on the gel (which may not actually be visible). *See* **Fig. 2**.

4. Notes

1. In practice, the range of PCR is determined by the incorporation of incorrect nucleotides into the extending strand. A newly generated

mismatch causes the polymerase to detach and the strand is then incomplete. This strand will therefore not provide a binding site for the second primer and is a "dead end" for amplification. The longer the product, the further the polymerase has to extend and the more likely it is that an incorrect incorporation will occur. Depending on the enzyme used a practical limit for PCR of <10 kB is observed. Several methods for extending the range of PCR have been developed that rely on including a low level of a thermostable poplymerase with proofreading activity into the enzyme mix. Mismatched nucleotides that are blocking extension are removed and the range is then substantially increased. Two minor concerns should be considered when using this approach:

   a. Proofreading enzymes can attack the 3' end of primers and reduce the specificity of binding, thereby producing an increased likelihood of false positives.
   b. In a plant where individual gene size (including introns) is generally quite small, such as *Arabidopsis*, increasing the range of detection may provide lines with inserts that flank the gene and do not disrupt it directly. Length-enhanced PCRs have, however, been very successfully used for reverse genetics in *Arabidopsis* (*7*).

2. In order to eliminate candidate inserts that may flank and not actually be in the gene, it can be useful to identify the actual insertion/disruption point of the mutagen. Sequencing of the diagnostic PCR product from individual lines using the insert primer (primer i; **Fig. 1**) can provide this if a sufficiently large region of known sequence belonging to the insert can be read. This approach is most useful if the native gene structure is partially or completely characterized.
3. In *Agrobacterium*-mediated transfer, the LB is not an absolute termination zone (***11***). The distance from the border to the insertion junction may therefore be extensive. In some cases, the LB itself will not be present in the insert. However, this does not present a large practical problem, because these individual inserts will not be detected in the screen using the LB primers. This will not substantially decrease the size of the population screened but may affect later analysis of the insert.
4. A very simple calculation would suggest that in *Arabidopsis*, which has an approx 100-mB genome (100,000 kB), a population containing 10,000 inserts is likely to have an insert every 10 kB if the inserts are randomly distributed. However, because this is a practical exercise in mathematical probability, a hit can never be guaranteed and it

quickly becomes prohibitively difficult to obtain the final 5% of the possible hits. This is a similar problem to that of estimating the likelihood of finding any given clone in screening a recombinant genomic DNA library *(12)*. Also, the assumption of random distribution of insertion may not necessarily be applicable to the chosen insert.

5. In our experience, using the widespread estimate that A/T contributes 2°C and G/C contributes 4°C toward calculating annealing temperatures *(13)* is usually as effective in practice as other more sophisticated methods or programs.
6. T-DNA insertions are often complex, in that they may comprise several conjoined inserts as partial or complete tandem duplicates, either as direct or inverted repeats *(14)*. This leads to the potential for LB-LB or RB-RB pairs from which PCR can initiate. The frequency of these complex insertions is very high in most T-DNA populations.
7. Tissue that has been frozen in liquid nitrogen can be crushed to reduce its volume, stored in a –20 or –70°C freezer for several years, and still remain suitable for extracting DNA for PCR. For storage we recommend 50-mL screw-top vials (Starstedt).
8. One frequent, but often unappreciated source of PCR contamination can be from tips or tubes that have been autoclaved. Autoclaving kills bacteria, but does not necessarily destroy DNA. Not only do autoclaves accumulate debris, but they may also splash and redistribute this material onto otherwise clean plasticware. Similarly, racking tips can lead to a false sense of security because handled tips can retain DNA contaminants throughout autoclaving. We recommend, therefore, that all tips and tubes are used fresh from the machine-packaged bag and are rigorously disposed of if accidentally touched. Similarly, although wearing gloves can help to reduce contamination, simply being aware of not touching potentially contaminating surfaces is much more important.
9. The buffer used may vary with the chosen polymerase (*see* **Note 1**) and with the primers (*see* **Note 11**).
10. As a positive control for the general quality and "PCR-ability" of the DNA, we amplify with a consistent primer pair designed to the *Sec Y* gene from *Arabidopsis* *(15)*. The primers we use (S. May, *unpublished data*) are described in **Table 1**. and give a strong 300-bp product if the DNA is usable.
11. We would not normally test a range of magnesium concentrations in our PCRs because we prefer to standardize around a common magnesium concentration and optimize the annealing temperature. In our

experience we have never found a primer set that did not work with standard buffer conditions. However, it can be time-saving to accommodate historical conditions (buffer, Mg^{2+}, cycle times) into a diagnostic PCR to match those used previously. New untested PCRs may be more conveniently served by adopting standard conditions and only varying the annealing temperature, because this is normally the single most critical variable.

12. A very large number of protocols are available for plant DNA extraction, most of which reliably produce DNA of sufficient purity to allow efficient PCR. The protocol given here is a fairly crude extraction that is perfectly adequate for most applications. If the DNA is intended for long-term storage, then we recommend use of dedicated kits (e.g., Qiagen) either in place of or subsequent to this step. For screening purposes, floral meristems are certainly the most suitable tissue to use since the plant will rapidly replace these. However, for large-scale extractions, leaves are likely to contain lower levels of polysaccharides that can interfere with extraction. DNA that has been stored after a crude extraction can be subject to browning and an associated loss in "PCR-ability" over time (even when frozen). To clean archival DNA of this type, it may be passed through a silica-based column, such as those used for commercial minipreps or gel extractions. Even dark-brown DNA may be recovered in this manner.
13. The standard volume of PCRs can be highly variable between labs. There is an argument that says that large volume PCRs, such as 50–100 µL, can require longer cycle times to effectively equilibrate core and wall temperatures, and can therefore be less reliable under some critical conditions. Small-volume PCRs (10–20 µL) are not prone to this problem, can easily produce high yields of product for cloning when efficient, and use fewer resources. We therefore normally use 10 µL PCRs for both diagnostic and cloning purposes. In hot lid conditions, where a covering evaporation shield (e.g., paraffin or mineral oil) is not used, small-volume PCRs (<10 µL) can, however, suffer from more significant volume drop than larger PCRs.
14. It is always preferable to make a megamix (nopri) comprising materials for a large number of PCRs mixed together such that all of the components are present except for the primers, template, and enzyme. The volume is calculated such that the missing components may be added in the combination required. This supports reproducibility across the PCRs. If a large number of PCRs are to be per-

formed that have identical primer pairs, it is also advisable to make suitable (+pri) megamixes that include these primers to ensure that conditions are as similar as possible for all equivalent PCRs.

15. In our experience, if PCRs do not work, then the primary candidate for improvement is almost always the inclusion of a hot start (*see* **Note 17**). Although it can make the manipulation of materials slightly more arduous, we cannot emphasize the importance of this as a trouble-shooting exercise if PCRs fail.
16. Cycle times and volumes of PCR are a contentious and sometimes highly personal issue. We have reliably and repeatedly used cycle times of: 15 s denature; 15 s anneal; 15 s extend to amplify PCR products as long as 5 kB. From many groups' results, it is clear that the denaturation time does not need to be longer, even for genomic DNA; the annealing time equally is quite sufficient, since the primers are in enormous molar excess in the reaction, and priming is therefore almost instantaneous; the only contentious item is the extension time (*see* **Note 18**), which may need to be increased for some types of proofreading polymerases (*see* **Note 1**). Because we realize that it can be difficult to persuade collaborators to use 15-s extension times, we have standardized on 60 s since this does not significantly delay total cycling times and is no less efficient.
17. At room temperature, the activity of thermostable polymerases is such that primers will bind at nonspecific sites and will be extended by the enzyme. These extended primers can then produce unwanted artefacts and templates for later amplification in the PCR. If, however, all of the components except one (the enzyme) are heated to the extension temperature (i.e., well above the specific annealing temperature of the primers) before addition of the final component, then artefacts can be drastically reduced or eliminated by this "hot start."
18. At 72°C, Taq polymerase extends approximately at a rate of 72 nucleotides/s; this normally leads to calculations of extending times that are kB/72 s. In practice, however, extension times of 15 s are sufficient to amplify fragments as long as 5 kB.
19. A grid size whereby each axis is the square root of the subpopulation size is most efficient, but not generally practical.
20. The advantages of digoxigenin as a label are principally speed, economy (because probes can be stored and reused), nonradioactive nature, and probably most important, consistency ***(16)***. Probes can be used and stored for a year or more in the –20°C freezer. For

reverse genetics, this can be essential, because the time between screens (especially if subpools are to be analyzed) can be days or even weeks if a generation has to be followed through. A dig probe does not suffer from decay and can therefore be used repeatedly (or aliquotted after labeling and stored to be used equivalently) over a period of months and probably years.

21. We advise the use of small area (mini-) gels in the diagnoses of PCRs by hybridization. This is not only economical in terms of membrane and hybridization materials, but it can also substantially increase the intensity of the signal by concentrating the same target band into a smaller area.

References

General *Arabidopsis* information can be obtained from The *Arabidopsis* Information Resource (TAIR) [http://www.arabidopsis.org].

1. Boguski, M. S., Lowe, T. M., and Tolstoshev, C. M. (1993) dbEST—database for "expressed sequence tags." *Nat. Genet.* **(4),** 332,333.
2. Schaefer, D. G. and Zryd, J. P. (1997) Efficient gene targeting in the moss physcomitrella patens. *Plant J.* **11,** 1195–1206.
3. Winkler, R. G. and Feldmann, K. A. (1998) PCR-based identification of T-DNA insertion mutants, in *Methods in Molecular Biology, vol. 82:* Arabidopsis *Protocols* (Martinez-Zapater, J. and Salinas, J., eds.), Humana, Totowa, NJ, pp. 129–136.
4. Walbot, V. (1992) Strategies for mutagenesis and gene cloning using transposon tagging and T-DNA insertional mutagenesis. *Ann. Rev. Plant. Physiol. Plant Mol. Biol.* **43,** 49–82.
5. Zambryski, P., Depicker, A., Kruger, K., and Goodman, H. M. (1982) Tumour induction by *Agrobacterium tumefaciens*: analysis of the RT boundaries of T-DNA. *J. Mol. Appl. Genet.* **1,** 361–370.
6. McKinney, E. C., Ali, N., Traut, A., Feldmann, K. A., Belostotsky, D. A., McDowell, J. M., and Meagher, R. B. (1995) Sequence-based identification of T-DNA insertion mutations in Arabidopsis: actin mutants act2-1 and act4-1. *Plant J.* **8(4),** 613–622.
7. Krysan, P. J., Young, J. C., Tax, F., and Sussman, M. R. (1996) Identification of transferred DNA insertions within *Arabidopsis* genes involved in signal transduction and ion transport. *Proc. Natl. Acad. Sci. USA* **93(15),** 8145–8150.

8. Fortsthoefel, N. R., Wu, Y., Schulz, B., Bennett, M. J., and Feldmann, K. A. (1992) T-DNA insertion mutagenesis in Arabidopsis: prospects and perspectives. Aust. *J. Plant Physiol.* **19,** 353–366.
9. Arabidopsis Stock Centres: Eurasia: NASC, University of Nottingham, University Park, Nottingham, NG7 2RD, UK, NASC on-line catalogue. <http://nasc. nott. ac. uk/>. The Americas: ABRC, The Ohio State University, 1735 Neil Avenue, 309 Botany and Zoology Bldg., Columbus, OH 43210, USA, ABRC on-line catalogue. <http://aims. cps. msu. edu/aims/>.
10. Ko, C. H., Brendel, V., Taylor, R. D., and Walbot, V. (1998) U-richness is a defining feature of plant introns and may function as an intron recognition signal in maize. *Plant Mol. Biol.* **36(4),** 573–583.
11. Koncz, C., Nemeth, K., Redei, G. P., and Schell, J. (1992) T-DNA insertional mutagenesis in Arabidopsis. *Plant Mol. Biol.* **20(5),** 963–976.
12. Old, R. W. and Primrose, S. B. (1996) Cloning Strategies, gene libraries, and cDNA cloning. In: *Principles of Gene Manipulation: An Introduction to Genetic Engineering*, 5th ed., Blackwell Scientific Publications, pp. 101,102.
13. Wu, D. Y., Ugozzoli, L., Pal, B. K., Qian, J., and Wallace, R. B. (1991) The effect of temperature and oligonucleotide primer length on the specificity and efficiency of amplification by the polymerase chain reaction. *DNA Cell Biol.* **10(3),** 233–238.
14. Feldmann, K. A. (1991) T-DNA insertion mutagenesis in Arabidopsis: mutational spectrum. *Plant J.* **1(1),** 71–82.
15. Laidler, V., Chaddock, A. M., Knott, T. G., Walker, D., and Robinson, C. (1995) A SecY homolog in *Arabidopsis thaliana*. Sequence of a full-length cDNA clone and import of the precursor protein into chloroplasts. *J. Biol. Chem.* **270(30),** 17,664–17,667.
16. May, S. T. (1998) Latin studies of plant architecture. *Lab. News* **576,** 20–22.
17. Sambrook, J., Fritsch, E. F., and Maniatis, T. (1989) *Molecular Cloning: A Laboratory Manual* Vol I–III 2nd ed., Cold Spring Harbor Laboratory Press, Cold Spring Harbor, NY.

Index

M

O

P

R

S

T

Z